LECTURE NOTES ON IMPEDANCE SPECTROSCOPY

Lecture Notes on Impedance Spectroscopy

Measurement, Modeling and Applications

Editor

Olfa Kanoun

Chair for Measurement and Sensor Technology
Technische Universität Chemnitz, Chemnitz, Germany

VOLUME 4

CRC Press
Taylor & Francis Group
Boca Raton London New York Leiden

CRC Press is an imprint of the
Taylor & Francis Group, an **informa** business
A BALKEMA BOOK

CRC Press/Balkema is an imprint of the Taylor & Francis Group, an informa business

Typeset by V Publishing Solutions Pvt Ltd., Chennai, India
Printed and bound in Great Britain by CPI Group (UK) Ltd, Croydon, CR0 4YY

Published by: CRC Press/Balkema
P.O. Box 11320, 2301 EH Leiden, The Netherlands
e-mail: Pub.NL@taylorandfrancis.com
www.crcpress.com – www.taylorandfrancis.com

ISBN: 978-1-138-00140-4 (Hbk)
ISBN: 978-1-315-79567-6 (eBook PDF)

Lecture Notes on Impedance Spectroscopy, Volume 4 – Kanoun (ed)
© 2014 Taylor & Francis Group, London, ISBN 978-1-138-00140-4

Table of contents

Lecture Notes on Impedance Spectroscopy, Volume 4 – Kanoun (ed)
© 2014 Taylor & Francis Group, London, ISBN 978-1-138-00140-4

Preface

Impedance Spectroscopy is a widely used and interesting measurement method applied in many fields such as electro chemistry, material science, biology and medicine. In spite of the apparently different scientific and application background in these fields, they share the same measurement method in a system identification approach and profit from the possibility to use complex impedance over a wide frequency range and giving interesting opportunities for separating effects, for accurate measurements and for simultaneous measurements of different and even non-accessible quantities.

For Electrochemical Impedance Spectroscopy (EIS) competency from several fields of science and technology is indispensable. Understanding electro chemical and physical phenomena is necessary for developing suitable models. Suitable measurement procedures should be developed taking the specific requirements of the application into account. Signal processing methods are very important for extracting target information by suitable mathematical methods and algorithms.

The scientific dialogue between specialists of Impedance Spectroscopy, dealing with different application fields, is therefore particularly important to promote the adequate use of this powerful measurement method in both laboratory and in embedded solutions.

The International Workshop on Impedance Spectroscopy (IWIS) has been established as a platform for promoting experience exchange and networking in the scientific and industrial field. It has been launched already in 2008 with the aim to serve for encouraging the sharing of experiences between scientists and to support new comers dealing with impedance spectroscopy. The workshop has been gaining increasingly more acceptance in both scientific and industrial fields and addressing not only more fundamentals, but also diverse application fields of impedance spectroscopy. By means of tutorials and special sessions, young scientist get a good overview of different fundamental sciences and technologies helping them to get expertize even in fields, which are not in the focus of their previous background.

This book is the fourth in the series Lecture Notes on Impedance Spectroscopy which has the aims to widen knowledge of scientists in this field and includes selected and extended contributions from the International Workshop on Impedance Spectroscopy (IWIS '12). The book reports about new advances and different approaches in dealing with impedance spectroscopy including theory, methods and applications. The book is interesting for researcher and developers in the field of impedance spectroscopy.

I thank all contributors for the interesting contributions and the reviewer who supported by the decision about publication with their valuable comments.

Prof. Dr.-Ing. Olfa Kanoun

Modeling and parameterization

Lecture Notes on Impedance Spectroscopy, Volume 4 – Kanoun (ed)
 ISBN 978-1-138-00140-4

Method for parameterization of impedance-based models with time domain data sets

Meike Slocinski
Daimler AG, Research and Advanced Engineering HV Battery Systems, Ulm, Germany

ABSTRACT: State determination of Li-ion cells is often accomplished with Electrochemical Impedance Spectroscopy (EIS). The measurement results are in frequency domain and used to describe the state of a Li-ion cell by parameterizing impedance-based models. Since EIS is a costly measurement method, an alternative method for the parameterization of impedance-based models with time-domain data easier to record is presented in this work. For this purpose the model equations from the impedance-based models are transformed from frequency domain into time domain. As an excitation signal a current step is applied. The resulting voltage step responses are the model equations in time domain. They are presented for lumped and derived for distributed electrical circuit elements, i.e. Warburg impedance, Constant Phase Element and RCPE. A resulting technique is the determination of the inner resistance from an impedance spectrum which is performed on measurement data.

Keywords: Electrochemical Impedance Spectroscopy, modeling, Li-ion cell, inner resistance, voltage step response, Constant Phase Element

1 INTRODUCTION

An emerging application for large quantities of Li-ion cells is the electrification of power trains, where Li-ion cells seem to be a proper technology which covers a broad spectrum of e-mobility applications like hybrid electric vehicles, plug-in hybrid electric vehicles and electric vehicles. For these applications a precise state determination of the Li-ion battery and its cells is mandatory to ensure a reliable operation of the vehicle.

In an laboratory environment, state determination of Li-ion cells is often accomplished with Electrochemical Impedance Spectroscopy (EIS). Using these impedance spectra, impedance-based models are parameterized in the frequency domain with well-known fitting procedures. These models are designed to reduce the dimension of the measurement data in order to describe the state of a Li-ion cell with few parameters. The parameters of these models can than be used to determine the cell's current state (SOx). There exists a variety of impedance-based models represented by electrical equivalent circuits.

Due to the high measurement and computational complexity as well as cost factors, frequency domain EIS measurements are not likely to be implemented on board in vehicles in the near future. An alternative approach is shown to parameterize impedance-based models with time domain data available on board, i.e. currents, battery or cell voltages and temperatures. Therefore, in this work, a method is proposed for the transformation of electrical circuit model equations from frequency domain into time domain model equations. Particularly for electrical circuit models containing distributed elements, e.g. Warburg impedances (WB), Constant Phase Elements (CPE), RCPE or ZARC elements, these transformations require fractional calculus methods, as will be presented in detail.

In order to prove the validity of this approach in a realistic scenario the transformation of an impedance spectrum into time domain for the determination of the conventionally defined inner resistance of a Li-ion cell will be shown. This allows for the comparison of the measurement results in frequency domain and time domain.

2 SYSTEMTHEORETICAL APPROACH

Figure 1 displays a systemtheoretical description of a Li-ion cell.

Impedance-based models are electrical equivalent circuits, containing lumped and distributed elements, e.g. resistors, capacitors, Warburg impedances. They are often used for modeling electrochemical systems like Li-ion cells [1] and describe the overvoltage η, see Figure 2. In this case, the current is the excitation and thus the input signal whereas the voltage is the response and thus the output signal of the system.

There exists a variety of impedance-based models differing in accuracy and number of model parameters, described by the parameter set $\vec{P}$. The model equation is a function of time or frequency and also a function of a parameter set $\vec{P}$.

The intention of the description of a measurement with an impedance-based model is the dimension reduction of the measurement data. The dimension reduced measurement is expressed by the parameter set $\vec{P}$ together with the model equation, for both frequency domain and time domain. The parameterization of the impedance-based model is performed according to the modeling procedure described in Section 2.3.

This work aims to describe the impedance-based models in time domain where the voltage response $u(t)$ is the model equation. The starting point is the system description in the frequency domain using the impedances $Z(s)$ as the model equation. Time domain model equations are thus derived analytically from the frequency domain using a specific excitation signal.

2.1 *System description in frequency domain*

The impedance $Z(s)$ represents the state of a Li-ion cell in frequency domain and can be illustrated by an impedance spectrum, e.g. in a Nyquist plot in the complex plane, see Figure 3. The impedance is the system function or the frequency response locus which is in case of a Li-ion cell the complex voltage $U(s)$, i.e. the system response, divided by the complex current $I(s)$, i.e. the excitation signal. In frequency domain $s = \sigma + j\omega$ is the complex frequency with a damping term σ and the angular frequency $\omega = 2\pi f$.

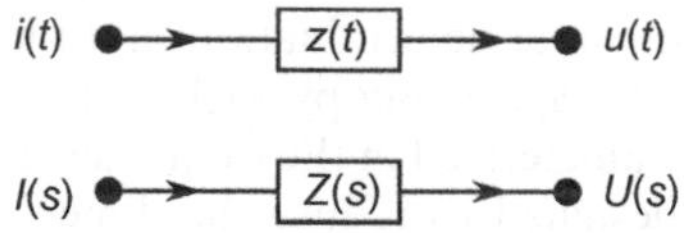
i(t)
z(t)
u(t)
I(s)
Z(s)
U(s)

Figure 1. Description of the system of a Li-ion cell or battery, which is a two-port network. The upper graph shows the system in time domain and the lower graph shows the system in frequency domain. The input signal is the current excitation and whereasthe output signal is the voltage response.

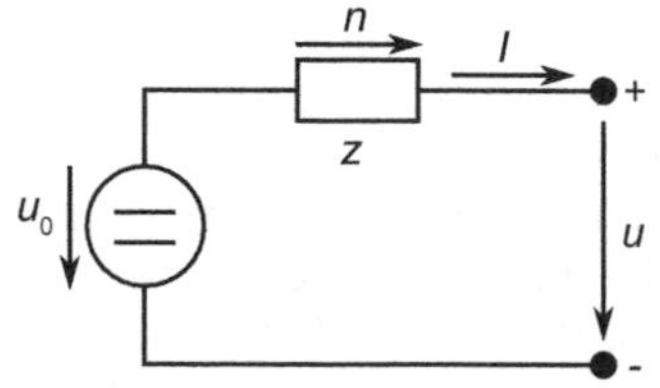
n
I
+
Z
u0
u
-

Figure 2. The graph shows the model of a Li-ion cell, with the open circuit voltage U_0, the closed circuit voltage U and the overvoltage η. The overvoltage is expressed by an impedance-based model with the impedance Z.

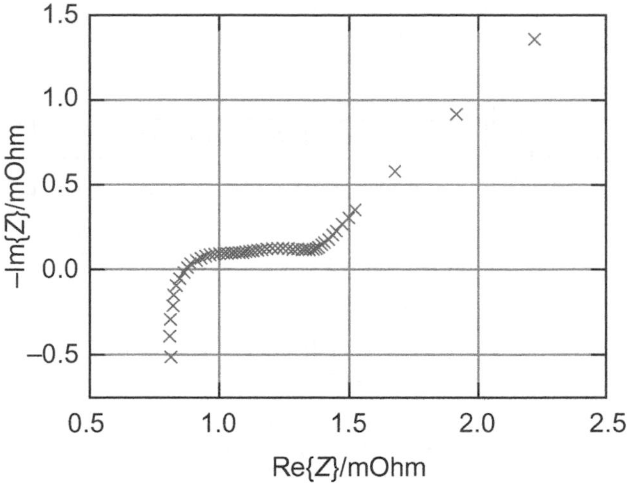
1.5
1.0
0.5
0.0
−0.5
−Im{Z}/mOhm
0.5
1.0
1.5
2.0
2.5
Re{Z}/mOhm

Figure 3. Example of an impedance spectrum of a Saft VL6P Li-ion cell.

A certain state of the system is described by the impedance $Z(s,\vec{P})$ as a function of a parameter set $\vec{P}$ which corresponds to the component values of the electrical equivalent circuit elements for that state, see Equation (1).

$$Z(s)=\frac{U(s)}{I(s)} \Rightarrow Z(s,\vec{P})=\frac{U(s,\vec{P})}{I(s)} \tag{1}$$

In order to determine the parameter set $\vec{P}$ for a certain state of the cell in the frequency domain, an impedance spectrum is recorded and modeled with $Z(s,\vec{P})$ according to Section 2.3.

2.2 *System description in time domain*

The model for a Li-ion cell in time domain is described by the output voltage $u(t)$ being the convolution of the input signal $i(t)$ and the system function $z(t)$. The voltage response $u(t,\vec{P})$ is a function of a certain parameter set $\vec{P}$, depending on the state of the Li-ion cell, see Equation (2).

$$u(t)=i(t)*z(t) \Rightarrow u(t,\vec{P})=i(t)*z(t,\vec{P}) \tag{2}$$

To determine the parameter set $\vec{P}$ in time domain, the voltage response $u(t,\vec{P})$ is modeled according to Section 2.3, assuming that the model equation considers the analytical description of the excitation signal $i(t)$ which is applied during the measurement.

2.3 *Parameterization in frequency and time domain*

In principle, the parameterization can be performed on frequency or time domain data, see Figure 4. But not all physical effects are covered in both data likewise, e.g. low temperatures and high currents, see Section 4.

From a mathematical point of view, the model equations are non-linear functions with several parameters, having a bounded range. The model equations are the impedance as a function of the complex frequency $Z(s)$ in frequency domain and the voltage response to a certain current excitation signal as a function of time $u(t)$ in time domain. In principle any

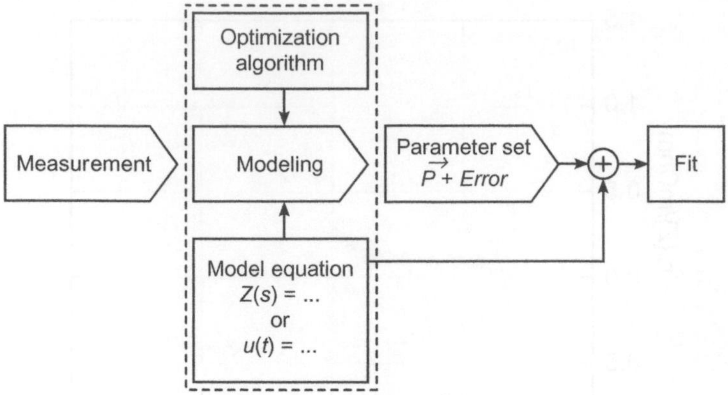
Optimization algorithm
Measurement
Modeling
Parameter set
$\vec{P}$ + Error
Fit
Model equation
Z(s) = ...
or
u(t) = ...

Figure 4. Process of modeling both time domain and frequency domain measurement data. The optimization process results in an adjusted parameter set $\vec{P}$ and an associated error. Together with the model equation the parameter set yields the fitting curve which is a reproduction of the measurement data.

optimization algorithm can be utilized for the modeling process, but it should be carefully selected in order to suit the mathematical problem. In this work parameterization of models is performed with the Particle Swarm Optimization (PSO) algorithm according to the modeling procedure described in [2].

The PSO is a nature inspired algorithm and allows to set boundary conditions for the model parameters and thus defines the search space, where the optimizations results are located likely. It is an iterative optimization process and during operation the search space is scanned for a minimum by a number of particles all representing possible solutions. These particles are influencing each other and share information about past and current positions in search space, i.e. the particle swarm shows intelligent behavior. The quality of an optimization result is quantified with a uniform distance measure. A detailed description of this metric can be found in [3].

The result of the modeling process is a set of parameters $\vec{P}$ that yields, together with the model equation used, a fitting curve of the measurement data together with an associated error.

3 MODEL TRANSFORMATION

If the electrical equivalent circuit model is a series connection of model elements, the overall model equation is the vector addition of the single model element equations in time domain and frequency domain, respectively. Time domain correspondences for single model elements to certain excitation signals can be calculated analytically as described in the following sections. In general these transformations can be easily computed for models consisting of lumped elements. For models containing distributed elements, fractional calculus is required.

For analytical calculations and easy metrological realization, the excitation signal should be a simple signal, e.g. Dirac delta function or the Heaviside step function. The metrological realization of the Dirac delta function is difficult, but it is easy to excite a Li-ion cell with a step-shaped current input signal. Current steps can even be found in real driving data or are easy to generate in laboratory. Furthermore, the corresponding time domain model equations can be calculated analytically.

In the following, common excitation signals are listed in time and frequency domain. Additionally the well-known voltage responses to current step excitations for single lumped elements (R, C, RC) are shown leading to the novel derivation for distributed elements

(WB, CPE, RCPE) using fractional calculus. Applying the single model element voltage step responses (VSR), the time domain model equations are set up for two common models.

3.1 *Excitation signals*

Common current excitation signals and their Laplace correspondences in frequency domain are listed in Table 1.

If the current excitation signal $i(t)$ and its Laplace correspondence $I(s)$ are known, the voltage response $u(t)$ to this excitation in time domain for a model element with an impedance $Z(s)$ can be calculated by inverse Laplace transform of the product of $I(s)$ and $Z(s)$, according to Equations (3) and (4).

$$U(s) = I(s) \cdot Z(s) \tag{3}$$

$$\bullet\!\!-\!\!\circ$$

$$u(t) = \mathcal{L}^{-1}\{I(s) \cdot Z(s)\} \tag{4}$$

3.2 *Derivation of model elements time domain correspondences with a current step excitation*

In the following current step excitations are used exclusively as excitation signals, due to their simple metrological realization and since it is possible to transform VSR in frequency domain into time domain analytically.

There exist two methods for the derivation of the transient time domain voltage responses $u(t)$. One approach is to solve the differential equations of $u(t)$ in time domain for the overall model equation derived according to Kirchhoff's laws. This method is appropriate for lumped elements and simple combinations of these elements.

As an alternative the single model elements' impedances $Z(s)$ connected in series in the model, are multiplied with the Laplace transform of the excitation signal $I(s)$. These products are the voltage responses $U(s)$ in frequency domain for the single model elements and are transformed into time domain using inverse Laplace transform subsequently. The overall voltage response of the model consists of the sum of voltage responses of the single model elements. This approach is used in the following.

For lumped elements, e.g. resistors, capacitors or combinations of these elements, the differential equations, impedances and VSR are well-known [4]. Distributed elements, i.e. Warburg impedance, Constant Phase Element, or parallel connections like RCPE, also known as ZARC or Cole-Cole element, have non-integer exponents α of the complex frequency s in frequency domain. This corresponds to fractional differential equations in time domain and thus the calculation of the VSR requires fractional calculus, as can be seen in the following derivations.

Table 1. Commonly used current excitation signals in time domain and their Laplace transforms in frequency domain. $\hat{I}$ is the amplitude of the excitation siganal.

Excitation signal	Time domain		Frequency domain
Arbitrary	$i(t)$	$\circ\!\!-\!\!\bullet$	$I(s)$
Dirac delta	$\hat{I} \cdot \delta(t)$	$\circ\!\!-\!\!\bullet$	$\hat{I} \cdot 1$
Heaviside step	$\hat{I} \cdot \sigma(t) = \hat{I} \begin{cases} 0, t < 0 \\ 1, t \geq 0 \end{cases}$	$\circ\!\!-\!\!\bullet$	$\hat{I} \cdot \frac{1}{s}$
Sine	$\hat{I} \cdot \sin(\omega t)$	$\circ\!\!-\!\!\bullet$	$\hat{I} \cdot \frac{\omega}{\omega^2 + s^2}$

a) *Constant Phase Element*

The underlying differential equation of the CPE is an arbitrary fractional derivative of the order α. The model parameters of the CPE are Q and α and a CPE shows capacitive behavior for $\alpha = (0, 1]$ and inductive behavior for $\alpha = [-1, 0)$. The VSR of a CPE in frequency domain is shown in Equation (5). Performing the inverse Laplace transform of $U_{\text{step, CPE}}(s)$ using the correspondence $\frac{\alpha!}{s^{\alpha+1}} \bullet\!\!-\!\!\circ\; t^{\alpha}$ [5] leads to the VSR in time domain of a CPE which includes an arbitrary-root shaped decay, see Equation (6).

$$U_{\text{step, CPE}}(s) = I(s) \cdot Z(s) = \frac{\hat{I}}{s} \cdot \frac{1}{Qs^{\alpha}} = \frac{\hat{I}}{Q\alpha!} \cdot \frac{\alpha!}{s^{\alpha+1}} \tag{5}$$

$$\bullet\!\!-\!\!\circ$$

$$u_{\text{step, CPE}}(t) = \frac{\hat{I}}{Q\alpha!} t^{\alpha} = \frac{\hat{I}}{Q} \cdot \frac{t^{\alpha}}{\Gamma(\alpha+1)} \tag{6}$$

With the amplitude of the excitation signal $\hat{I}$, $\alpha! = \Gamma(\alpha+1)$ and the Gamma function $\Gamma(x)$ which is an extension of the factorial function to real and complex numbers, see [6].

b) *Warburg impedance*

The Warburg impedance is a special case of a CPE with the constant exponent $\alpha = 0.5$, i.e. the underlying differential equation contains the half fractional derivative. Thus the VSR contains a square-root shaped decay according to Equation (7).

$$u_{\text{step, WB}}(t) = \frac{\hat{I}}{Q} \cdot \frac{\sqrt{t}}{\Gamma(1.5)} \tag{7}$$

With the amplitude of the excitation signal $\hat{I}$, the Parameter of the Warburg impedance Q and the constant value $\Gamma(1.5) \approx 0.88623$.

c) *RCPE*

The RCPE is a parallel connection of a resistor and a CPE and thus described by the model parameters R, Q and α. Its VSR in frequency domain is shown in Equation (8).

$$U_{\text{step, RCPE}}(s) = I(s) \cdot Z(s) = \frac{\hat{I}}{s} \cdot \frac{R}{1 + RQs^{\alpha}} \tag{8}$$

In order to derive the VSR of the RCPE $u_{\text{step, RCPE}}(t)$ in time domain, the inverse Laplace transform of Equation (8) has to be derived using fractional calculus, see Equations (9)–(16).

For the special case $Z(s) = \frac{1}{as^{\alpha}+b}$ the VSR $u_{\text{step}}(t)$ is shown in Equation (9) according to [7].

$$u_{\text{step}}(t) = D^{-1}z(t) = \frac{1}{a}\varepsilon_0\left(t, -\tfrac{b}{a}; \alpha, \alpha+1\right) \tag{9}$$

With the Mittag-Leffler function $E_{\alpha,\beta}(x)$ as defined in Equation (10), [7].

$$E_{\alpha,\beta}(x) = \sum_{n=0}^{\infty} \frac{x^n}{\Gamma(\alpha n + \beta)} \quad (\alpha > 0, \beta > 0) \tag{10}$$

and the k-th derivative of the Mittag-Leffler function as defined in Equation (11), [7].

$$E_{\alpha,\beta}^{(k)}(x)=\sum_{n=0}^{\infty}\frac{(n+k)!x^n}{n!\Gamma(\alpha n+\alpha k+\beta)} \qquad (k=0,1,2,...) \tag{11}$$

Thus the function ε_k can be defined as in Equation (12), [7].

$$\varepsilon_k(t,y;\alpha,\beta)=t^{\alpha k+\beta-1}E_{\alpha,\beta}^{(k)}\left(yt^{\alpha}\right) \qquad (k=0,1,2,...) \tag{12}$$

And the k-th derivative $D^{\lambda}\varepsilon_k$ of the function ε_k is defined in Equation (13). With D^{λ} being the fractional differential operator to the order λ, [7].

$$D^{\lambda}\varepsilon_k(t,y;\alpha,\beta)=\varepsilon_k(t,y;\alpha,\beta-\lambda) \qquad (\lambda<\beta) \tag{13}$$

In Equation (9) D^{-1} is the first negative derivative, i.e. the first positive integral.
The VSR can thus be rewritten from Equation (9):

$$u_{step}(t)=\frac{1}{a}\varepsilon_0\left(t,-\tfrac{b}{a};\alpha,\alpha+1\right)=\frac{1}{a}t^{\alpha}E_{\alpha,\alpha+1}\left(-\tfrac{b}{a}t^{\alpha}\right) \tag{14}$$

$$=\frac{1}{a}t^{\alpha}\sum_{n=0}^{\infty}\frac{\left(-\frac{b}{a}\right)^n t^{\alpha n}}{\Gamma(\alpha n+\alpha+1)} \tag{15}$$

$$=\frac{1}{a}t^{\alpha}\sum_{n=0}^{\infty}\left(-\frac{b}{a}\right)^n\frac{t^{\alpha n}}{(\alpha n+\alpha)!} \tag{16}$$

For the case of a RCPE $a=RQ$ and $b=1$. Thus the VSR contains an $\frac{1}{\alpha}$-root-shaped voltage response and a Mittag-Leffler function shaped voltage response, according to Equation (18).

Table 2. The table shows for often used elements for impedance based models the differential equations in time domain, the impedances and the VSR. The overall impedance of a model containing these elements can be set up by summing up the equations of the single model elements. $\Gamma(x)$ denotes the Gamma function and $E_{\alpha,\beta}(x)$ denotes the Mittag-Leffer function.

Model element	Differential equation $i(t)$ =	Impedance $Z(s)$ =	Voltage step response $u_{step}(t)$ =
R	$\frac{u(t)}{R}$	R	$\hat{I}R$
C	$C\frac{d}{dt}u(t)$	$\frac{1}{Cs}$	$\frac{\hat{I}}{C}t$
RC	$\frac{u(t)}{R}+C\frac{d}{dt}u(t)$	$\frac{R}{1+RCs}$	$\hat{I}R\left(1-\exp\left(-\frac{t}{RC}\right)\right)$
WB	$Q\frac{d^{0.5}}{dt^{0.5}}u(t)$	$\frac{1}{Qs^{0.5}}$	$\frac{\hat{I}}{Q}\frac{t^{0.5}}{\Gamma(1.5)}$
CPE	$Q\frac{d^{\alpha}}{dt^{\alpha}}u(t)$	$\frac{1}{Qs^{\alpha}}$	$\frac{\hat{I}}{Q}\frac{t^{\alpha}}{\Gamma(\alpha+1)}$
RCPE	$\frac{u(t)}{R}+Q\frac{d^{\alpha}}{dt^{\alpha}}u(t)$	$\frac{R}{1+RQs^{\alpha}}$	$\frac{\hat{I}}{Q}t^{\alpha}E_{\alpha,\alpha+1}\left(-\frac{1}{RQ}t^{\alpha}\right)$

$$U_{step,\mathrm{RCPE}}(s) = \hat{I}R\frac{1}{s}\frac{1}{1+RQs^{\alpha}} \tag{17}$$

$$\updownarrow$$

$$u_{setp,\,RCPE}(t) = \frac{\hat{I}}{Q}t^{\alpha}\sum_{n=0}^{\infty}\left(-\frac{1}{RQ}\right)^{n}\frac{t^{\alpha n}}{(\alpha n+\alpha)!} \tag{18}$$

This result conforms with the VSR of a RCPE presented in [8].

For commonly used lumped and distributed elements, the differential equations, impedances and VSR are summarized in Table 2.

3.3 *Examples for models*

Many impedance-based models have been proposed in the past [9–13]. It turned out that the models depicted in Figure 5 and 6 are able to match the measured impedance spectra to a high degree [2]. In the following, both models are presented with their characteristic properties, their impedances and VSR, which are derived according to the method proposed in the previous sections.

a) *R-RC-RC-WB model*

This model contains lumped circuit elements and one Warburg impedance, and its parameter set has six elements $\vec{P} = \left[R_2, R_3, R_4, C_3, C_4, Q_5\right]$. In the Nyquist plot this model results in two symmetric semicircles and a –45° diffusion branch shifted on real axis with the value of the series resistance R_2. Figure 5 shows the electrical equivalent circuit and the impedance is given by Equation (19).

$$Z(s) = R_2 + \frac{R_3}{1+R_3C_3s} + \frac{R_4}{1+R_4C_4s} + \frac{1}{Q_5s^{0.5}} \tag{19}$$

The VSR in time domain includes exponential function decays and a square root decay caused by the Warburg impedance, see Equation (21). This model is a good compromise between fitting accuracy and computational effort.

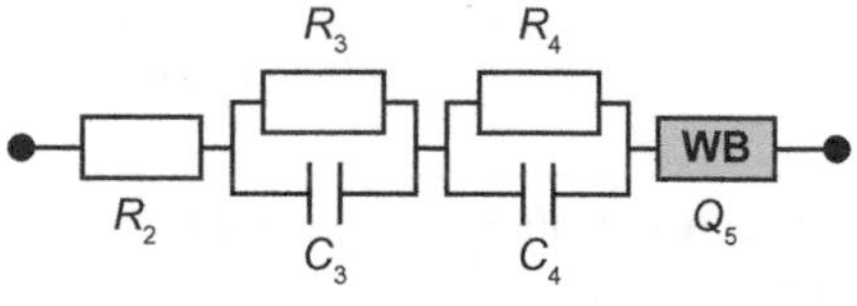
R3
R4
WB
R2
C3
C4
Q5

Figure 5. The figure shows the R-RC-RC-WB model. Besides the Warburg impedance, this model contains only lumped elements. Its impedance is shown in Equation (19), the VSR in frequency domain is shown in Equation (20) and the VSR in time domain is shown in Equation (21).

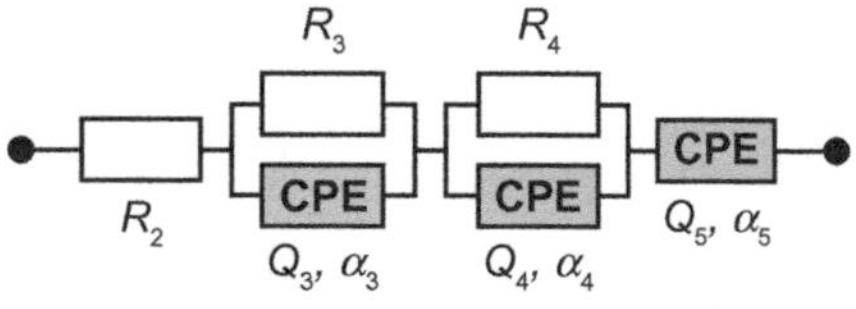
R3
R4
CPE
CPE
CPE
R2
Q3, α3
Q4, α4
Q5, α5

Figure 6. The figure shows the R-RCPE-RCPE-CPE model. This model contains only distributed elements besides the series resistance R_2. Its impedance is shown in Equation (22), the VSR in frequency domain is shown in Equation (23) and the VSR in time domain is shown in Equation (24).

$$U_{\text{step}}(s) = \hat{I}\left[\frac{R_2}{s} + \frac{R_3}{s + R_3 C_3 s^2} + \frac{R_4}{s + R_4 C_4 s^2} + \frac{1}{Q_5 s^{1.5}}\right] \tag{20}$$

$$\bullet\!\!-\!\!\circ$$

$$u_{\text{step}}(t) = \hat{I}\left[R_2 + R_3\left(1 - e^{-\frac{t}{R_3 C_3}}\right) + R_4\left(1 - e^{-\frac{t}{R_4 C_4}}\right) + \frac{\sqrt{t}}{Q_5 \Gamma(1.5)}\right] \tag{21}$$

b) *R-RCPE-RCPE-CPE model*

This model contains two RCPE and one CPE and therefore has nine parameters $\vec{P} = [R_2, R_3, R_4, Q_3, Q_4, Q_5, \alpha_3, \alpha_4, \alpha_5]$. In the Nyquist plot this model corresponds to two depressed semicircles and an arbitrary angled diffusion branch shifted on the real axis with the value of the series resistance R_2. Figure 6 shows the electrical equivalent circuit and the impedance is given by Equation (22).

$$Z(s) = R_2 + \frac{R_3}{1 + R_3 Q_3 s^{\alpha_3}} + \frac{R_4}{1 + R_4 Q_4 s^{\alpha_4}} + \frac{1}{Q_5 s^{\alpha_5}} \tag{22}$$

The VSR in time domain includes Mittag-Leffler function decays and an arbitrary root-shaped decay caused by the single CPE element, see Equation (24). For the calculation of the Mittag-Leffler functions, there exist toolboxes for MATLAB. This model provides an excellent fitting accuracy of the impedance spectra, but the parameterization in time domain requires more computational effort due to the number of parameters and the complexity of the Mittag-Leffler functions.

$$U_{\text{step}}(s) = \hat{I}\left[\frac{R_2}{s} + \frac{R_3}{s + R_3 Q_3 s^{\alpha_3+1}} + \frac{R_4}{s + R_4 Q_4 s^{\alpha_4+1}} + \frac{1}{Q_5 s^{\alpha_5+1}}\right] \tag{23}$$

$$\bullet\!\!-\!\!\circ$$

$$u_{\text{step}}(t) = \hat{I}\left[R_2 + \frac{1}{Q_3} t^{\alpha_3} E_{\alpha_3,\, \alpha_3+1}\left(-\tfrac{1}{R_3 Q_3} t^{\alpha_3}\right) + \frac{1}{Q_4} t^{\alpha_4} E_{\alpha_4, \alpha_4+1}\left(-\tfrac{1}{R_4 Q_4} t^{\alpha_4}\right) + \frac{1}{Q_5}\frac{t^{\alpha_5}}{\Gamma(\alpha_5 + 1)}\right] \tag{24}$$

4 EXPERIMENTAL

A proof and a possible application for this method is the determination of the inner resistance R_i from an impedance spectrum. This technique enables comparisons between measured impedance spectra and conventionally determined inner resistances.

The inner resistance of a Li-ion cell is defined as the voltage drop after a certain time t_x due to a current step excitation $i(t) = \hat{I} \cdot \sigma(t)$ with the amplitude $\hat{I}$. Thus the inner resistance is a time-dependent value, see Equation (25).

$$R_i(t_x) = \frac{u(t_x) - u(t_0)}{\hat{I}} \tag{25}$$

To determine the inner resistance from an impedance spectrum, the EIS measurement is carried out and parameterized in frequency domain with an impedance-based model, leading to the parameter set $\vec{P}$. The model is transformed analytically from frequency domain into time domain under the assumption of a step-shaped current excitation signal. The parameter set determined from the impedance spectrum is inserted into the analytically derived time domain VSR equation leading to the theoretical VSR. This VSR is evaluated for the times t_0 and t_x and the inner resistance is calculated according to Equation (25), see Figure 7.

4.1 *Measurements*

Measurements were performed on three Saft VL6P Li-ion cells with a nominal capacity of 7 Ah at their begin of life (BOL). For 13 state of charge (SOC) steps, seven temperatures between −15°C and 50°C, impedance spectra (51 frequencies, 10 mHz…5 kHz) as well as voltage responses due to current step excitation with different amplitudes, in discharging as well as in charging mode were carried out.

For the same state (SOC, temperature and age) of a Li-ion cell the impedance spectra and the VSR are compared, by computing the VSR from the impedance spectrum. Additionally, the inner resistances are computed from the determined and measured VSR. For the modeling process the proposed R-RC-RC-WB model was used.

Figure 8 shows the modeling results of the impedance spectra at BOL, 25°C, 100% SOC, modeled with the R-RC-RC-WB model and PSO algorithm described in [2]. Since this model includes no inductance, the inductive part in the positive imaginary semi-axis can not be fitted. Table 3 shows the values of the modeled parameter set.

Inserting the modeled parameter set into the analytically determined VSR of the R-RC-RC-WB model, Equation (21), leads to the theoretically determined VSR. Figure 9 shows the comparison of the measured VSR as well as the VSR determined from the impedance spectrum for different amplitudes $\hat{I}$ but the same state. The determined VSR fit the measured VSR very well and are mainly lying within the noise range of the measurement.

Inner resistances are evaluated and compared both, from the measured VSR $R_{i,\text{TD}}$ and from the impedance spectra determined VSR $R_{i,\text{FD}}$, at different times t_x and for different current excitation amplitudes $\hat{I}$, see Table 4. Both kinds of inner resistances agree very well

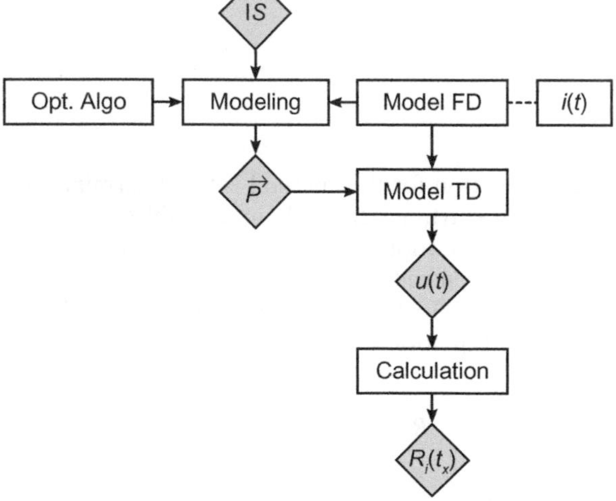
IS
Opt. Algo
Modeling
Model FD
i(t)
P
Model TD
u(t)
Calculation
R_i(t_x)

Figure 7. Process for the determination of the inner resistance $R_i(t_x)$ from an impedance spectrum (IS).

and their differences as well as the negative values of the $R_{i,\ TD}$ are due to the noisy measurement data.

For low temperatures and high excitation current amplitudes deviations occur, so that the voltage drop determined from the impedance spectrum is larger than the voltage drop measured, see Figure 10. In the impedance-based model used the influence of the current amplitude is only considered by the constant scaling factor $\hat{I}$. Non-linear influences of the current, e.g. Butler-Volmer non-linearities, are not included in the model. Furthermore, during the current step excitation with high current amplitudes the temperature of the cell increases and the SOC might change significantly, which is also not taken into account in the model.

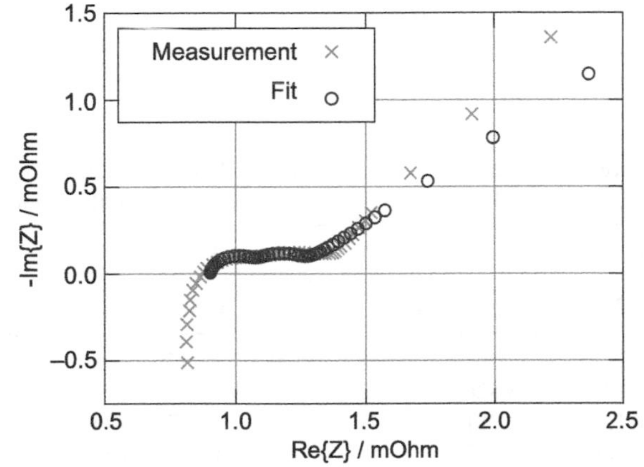

Measurement ×
Fit ○
1.5
1.0
0.5
0.0
−0.5
-Im{Z} / mOhm
0.5
1.0
1.5
2.0
2.5
Re{Z} / mOhm

Figure 8. Measured and modeled impedance spectrum of a Saft VL6P Li-ion cell at BOL, 25°C, 100% SOC, modeled with the R-RC-RC-WB model and PSO algorithm. For the modeled parameter set see Table 3.

Table 3. Modeled parameter set for the R-RC-RC-WB model of a Saft VL6P Li-ion cell at BOL, 25°C, 100% SOC. For the Nyquist plot of the measured and modeled impedance spectra see Figure 8.

Model parameter	**Parameter value**
R_2	$9.01 \cdot 10^{-4}\ \Omega$
R_3	$1.64 \cdot 10^{-4}\ \Omega$
R_4	$1.48 \cdot 10^{-4}\ \Omega$
C_3	$5.98 \cdot 10^{0}$ F
C_4	$9.00 \cdot 10^{1}$ F
Q_5	$2.46 \cdot 10^{3}\ S\sqrt{s}$

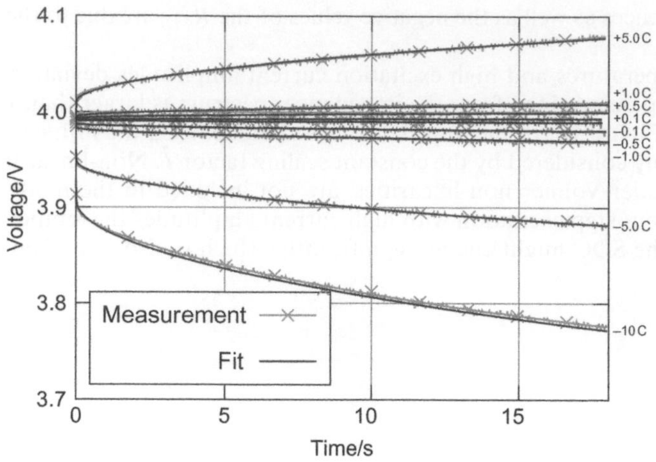

Voltage/V
Time/s
Measurement
Fit
+5.0 C
+1.0 C
+0.5 C
+0.1 C
−0.1 C
−0.5 C
−1.0 C
−5.0 C
−10 C

Figure 9. Comparison of the measured VSR (measurement) as well as the VSR determined from the impedance spectrum (fit) for different amplitudes $\hat{I}$ but the same state of a Saft VL6P Li-ion cell (BOL, 25°C, 100% SOC).

Table 4. Inner resistances determined from the measured VSR $R_{i,\,TD}$ and determined from the modeled impedance spectra $R_{i,\,FD}$, at different times t_x and for different current excitation amplitudes $\hat{I}$.

	$R_{i,\text{TD}}$/mΩ					$R_{i,\text{FD}}$/mΩ				
$\hat{I}$/C – rate	0.5 s	2 s	5 s	10 s	18 s	0.5 s	2 s	5 s	10 s	18 s
−0.1 C	1.719	2.210	−2.701	−2.456	3.684	1.537	1.861	2.238	2.663	3.159
+0.1 C	1.864	1.864	3.826	4.317	3.826	1.537	1.861	2.238	2.663	3.159
−0.5 C	1.568	1.861	2.351	2.841	3.429	1.537	1.861	2.238	2.663	3.159
+0.5 C	1.685	1.979	3.057	2.959	3.547	1.537	1.861	2.238	2.663	3.159
−1.0 C	1.543	1.886	1.739	2.719	3.306	1.537	1.861	2.238	2.663	3.159
+1.0 C	1.577	1.871	2.704	2.802	3.390	1.537	1.861	2.238	2.663	3.159
−5.0 C	1.567	1.861	2.292	2.674	3.154	1.537	1.861	2.238	2.663	3.159
+5.0 C	1.571	1.855	2.169	2.639	3.207	1.537	1.861	2.238	2.663	3.159
−10.0 C	1.539	1.818	2.171	2.563	3.101	1.537	1.861	2.238	2.663	3.159

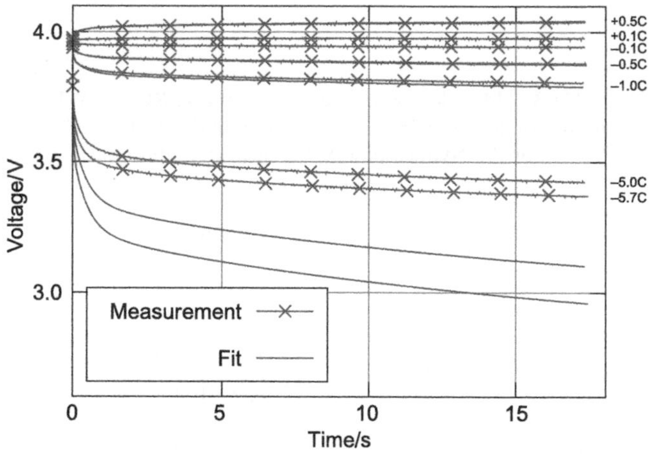

Voltage/V
Time/s
Measurement
Fit
+0.5C
+0.1C
−0.1C
−0.5C
−1.0C
−5.0C
−5.7C

Figure 10. Comparison of the measured VSR (measurement) as well as the VSR determined from the impedance spectrum (fit) for different amplitudes $\hat{I}$ but the same state of a Saft VL6P Li-ion cell (BOL, −15°C, 100% SOC).

5 CONCLUSION AND OUTLOOK

A method was proposed for the parameterization of impedance based models in the time domain, by deriving the corresponding time domain model equation with inverse Laplace transform of the frequency domain model equation assuming a current step excitation. This excitation signal has been chosen, since it can be easily applied to a Li-ion cell in an experiment, allows the analytical calculation of the time domain model equation and is included in the definition of the inner resistance. The voltage step responses of model elements were presented for lumped elements and derived for distributed model elements that have underlying fractional differential equations using fractional calculus. The determination of the inner resistance from an impedance spectrum was proposed as a possible application for this method. Tests on measurement data showed that this method works well for temperatures around room temperature and current excitation amplitudes up to 10 C. This technique can be used for comparisons of measured impedance spectra with conventionally determined inner resistances.

Another application is the parameterization of models using time series data from cells recorded during real operation of the electrified vehicle. Therefore the recorded data is scanned for current steps, e.g. with pattern recognition methods, and an impedance-based model is parameterized with the methods described. Furthermore, the model parameters can be utilized for determining an arbitrary defined SOx as a function of the modeled parameter set, e.g. the simple definition of the conventional inner resistance. Besides measurement data is smoothed during the modeling process eliminating noise and peaks. So preconditioning is no longer necessary when evaluating noisy data. The model parameters obtained from both, time domain and frequency domain models, can be used for simulation tasks, by integrating them as look-up tables for the state description.

The models used have their limits at low temperatures and high currents, since non-linearities concerning the current and changes in SOC and temperature during the current step excitation are not considered in this model. Therefore the expected voltage step responses are larger under these conditions than the voltage responses measured. Furthermore, the Li-ion cells have to be excited with current excitation signals the time domain model is including, what might be depending on the excitations signal's shape a challenging metrological task. For good fitting accuracy the sampling rates of the measurement have to ensure to detect small changes in the voltage responses.

For future work it is eligible to analyze the dependencies like high currents and low temperatures, to integrate them into the models and to carry out quantitative evaluations of the accuracy of the models. Furthermore, the technique should be applied the other way around to derive impedance spectra from voltage step response data. This enables comparisons of model parameter sets from time domain and frequency domain data. In order to prove state determination and tracking of the aging process, the techniques will be applied to a large number of recorded data from real operation.

ACKNOWLEDGMENT

The author would like to thank Stefan Richter and Philipp Wursthorn from DeutscheACCUmotive GmbH & Co. KG for carrying out the measurements this method was validated on.

REFERENCES

[1] E. Barsoukov and J.R. Macdonald, *Impedance Spectroscopy*. John Wiley & Sons, 2005.

[2] J. Roth, M. Slocinski, and J.-F. Luy, "Modeling li-ion battery aging data utilizing particle swarm optimization," in *Lecture Notes on Impedance Spectroscopy II* (O. Kanoun, ed.), pp. 1–11, CRC Press, 2011.

[3] M. Slocinski, K. Kögel, and J.-F. Luy, "Distance measure for impedance spectra for quantified evaluations," in *Lecture Notes on Impedance Spectroscopy III* (O. Kanoun, ed.), pp. 1–11, CRC Press, 2012.
[4] T. Frey and M. Bossert, *Signal- und Systemtheorie*. B.G. Teubner Verlag, 2004.
[5] H. Bateman, *Tables of integral transforms*. McGraw-Hill, 1954.
[6] I.N. Bronstein, K.A. Semendjajew, G. Musiol, and H. Mühlig, *Taschenbuch der Mathematik*. Frankfurt am Main: Verlag Harri Deutsch, 6. ed., 2005.
[7] I. Podlubny, *Fractional Differential Equations*. Academic Press, 1999.
[8] A. Salkind, T. Atwater, P. Singh, S. Nelatury, S. Damodar, C. Fennie Jr., and D. Reisner, "Dynamic characterisation of small lead-acid cells," *Journal of Power Sources*, vol. 96, pp. 151–159, 2001.
[9] D. Andre, M. Meiler, K. Steiner, H. Walz, T. Soczka-Guth, and S.D.U., "Characterization of high-power lithium-ion batteries by electrochemical impedance spectroscopy. ii:modelling," *Journal of Power Sources*, vol. 196, pp. 5349–5356, 2011.
[10] S. Buller, M. Thele, R. De Doncker, and E. Karden, "Impedance-based simulation models of supercapacitors and li-ion batteries for power electronic applications," *IEEE Transactions on Industry Applications*, vol. 41, no. 3, pp. 742–747, 2005.
[11] M. Danzer and E. Hofer, "Electrochemical parameter identification—an efficient method for fuel cell impedance characterisation," *Journal of Power Sources*, vol. 183, pp. 55–61, 2008.
[12] M. Einhorn, V. Conte, C. Kral, and J. Fleig, "Comparison of electrical battery models using a numerically optimized parameterization method," in *Vehicle Power and Propulsion Conference (VPPC), 2011 IEEE*, pp. 1–7, sept. 2011.
[13] J.B. Gerschler, J. Kowal, M. Sander, and D.U. Sauer, "High-spatial impedance-based modeling of electrical and thermal bahaviour of lithium-ion batteries," in *Electric Vehicle Symposium (EVS 23), Anaheim*, 2007.

Lecture Notes on Impedance Spectroscopy, Volume 4 – Kanoun (ed)

A nonlinear impedance standard

H. Nordmann
Electrochemical Energy Conversion and Storage Group, Institute for Power Electronics and Electrical Drives (ISEA), RWTH Aachen University, Germany
Jülich Aachen Research Alliance, JARA-Energy, Germany

D.U. Sauer
Electrochemical Energy Conversion and Storage Group, Institute for Power Electronics and Electrical Drives (ISEA), RWTH Aachen University, Germany
Institute for Power Generation and Storage Systems (PGS), E.ON ERC, RWTH Aachen University, Germany
Jülich Aachen Research Alliance, JARA-Energy, Germany

M. Kiel
Leopold KOSTAL GmbH & Co., KG Lüdenscheid, Germany

ABSTRACT: Since Electrochemical Impedance Spectroscopy has become more and more important not only in basic research, but also in application engineering, calibration and validation of impedance results gained from different measurement devices are among the major challenges in impedance spectroscopy. Since most electrochemical systems under investigation are not long-term stable, the validation and recalibration of the measurement equipment is not possible using an electrochemical cell. The nonlinear impedance standard described here is an electronic device which reproduces a typical impedance spectrum of a battery by using various passive and active electronic parts. This work points out which issues have to be addressed when developing such a device, followed by test results, ending with a short discussion and the upcoming improvements based on the last results.

Keywords: Impedance standard, nonlinearity, simulation, four point measurment, HELIOS

1 INTRODUCTION

For the comparison between various measuring methods and various test equipment for Electrochemical Impedance Spectroscopy (EIS) on batteries a reference is needed to make sure the battery test results of these different test equipment are comparable. The reference is not supposed to be changing its behavior between individual measurements. For calibration, an impedance standard for impedance spectroscopy is very important as well, because especially for the imaginary part there are no calibrated references available. Currently, calibration is done with a high precision shunt, which has a defined real part, but no imaginary part. Thus, the results measured by various EIS equipment are not easy to be compared. Besides the dependence of temperature and superimposed DC current, the impedance spectrum of a battery also depends on the SOC. Furthermore the battery's aging has an influencing role on the impedance spectrum as well [1]. Therefore, batteries or electrochemical cells are not suitable to make an impedance standard. The impedance standard is motivated by the HELIOS project's need for comparable results on aging tests. EIS measurements were performed, because they are valuable to gain knowledge upon aging processes, how different materials cope with aging procedures and of course the parameterization of aging models [2]. Five different test laboratories, each having different equipment, were involved to test high energy lithium batteries and perform EIS measurements after they were cycled or stored under rapid aging conditions. To make sure, all test labs have the same and comparable results, a method had to be implemented to make sure the test labs' measuring method is consistent enough to later draw valuable conclusions from the actual test results from each individual test lab.

In the following an electrical circuit is described which has a battery-like behavior in the frequency band between 1 mHz and 10 kHz, but does not show any variation in spectrum after extended storage or transport. At the end there is a discussion about adding a nonlinearity regarding current and voltage behavior which emulates the Butler-Volmer characteristics.

The gain crossover frequency $f_{\pm}$, where the imaginary part becomes zero, is found in the range between 100 Hz to 10 kHz. The impedance behavior of a battery in the lower frequency range can be described by several R–R||C elements. The main issue at reproducing a battery's impedance behavior is the capacities of the equivalent circuit. The capacity values being present in those elements can be up to several hundred Farads [3]. The realization of capacities with this size is possible either by a significant financial effort by paralleling electrolyte capacitors or using double layer capacitors. Unfortunately the issue with these capacitors is that their behavior depends on temperature, SOC and aging [4]. Also, double layer capacitors are not suitable because of their large size. A different approach utilizes the transformation of the impedance by a resistor network. This method is discussed in the following and is based on the demonstration of [5]. This approach was expanded for investigations regarding current dependency and a small size that is cost effective at the same time [6].

2 DESIGN

The principle which is used by the impedance standard (reference impedance) is based on the simple idea to form the excitation current into a voltage signal, which is then fed through a filter into the impedance spectroscope's measuring input. The Bode plot is then reproduced in the impedance spectrum.

2.1 *Circuit diagram*

The principle circuit diagram of the reference impedance is shown in Figure 1.

The excitation signal from the power amplifier is fed into the shunt resistor Rshunt and transformed into a voltage signal. This voltage signal is then fed into a voltage divider that has a complex impedance element. The measured impedance $\underline{Z}_g$ is given by (1).

$$\underline{U} = \frac{\underline{Z}}{\underline{Z} + R_v + R_{shunt}} \cdot \underline{I} \cdot R_{shunt} \rightarrow \underline{Z}_g = \frac{\underline{U}}{\underline{I}} = \frac{\underline{Z} \cdot R_{shunt}}{\underline{Z} + R_v + R_{shunt}} \quad (1)$$

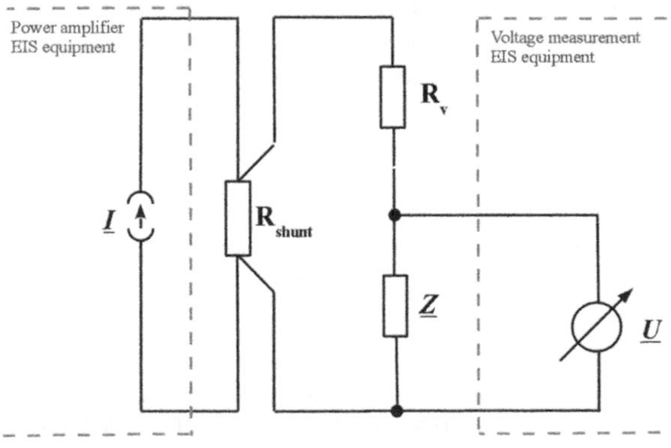

Power amplifier
EIS equipment
Voltage measurement
EIS equipment
R_v
I
R_shunt
Z
U

Figure 1. Circuit diagram of the reference impedance.

The impedance $\underline{Z}$ can be built from discrete parts, even though their individual impedance is relatively large for the typical frequency range of 10 mHz to 10 kHz. As one can observe from (1) the impedance $\underline{Z}$ is transformed by R_{shunt} and R_v. The key feature is having the voltage response measured not at the same point where the EIS meter's power amplifier is usually connected to during an EIS measurement. With this decoupling method the transfer function between input (power amplifier) and output (voltage response) can be set by designing proper elements for it.

2.2 *Calculation of needed elements*

The following Figure 2 shows how to lay out an impedance standard. The figure shows a simple R–R||C circuit which forms Z. The transfer locus draws a semicircle into the Nyquist plot. Its first intersection point at high frequencies gives the value of R_s, whereas the second intersection point with the real axis gives the sum of R_s and R_p. R_s, R_p and C_p make the complex impedance Z, which is going to be transformed into the impedance $\underline{Z}_m = \frac{U}{I}$ by the resistor R_v and the shunt resistor R_{shunt}. The shunt resistor is to be chosen small enough to not cause excessive heat and also to not over stress the power amplifier in the impedance meter. Besides the already mentioned considerations, the value for R_{shunt} must be obtained by also taking into account, that the voltage signal level that is fed into R_v and $\underline{Z}$ needs to be large enough. In the following example a value of 1Ω for R_{shunt} was chosen.

The ratio of R_v and R_{shunt} must be large enough so that the cross current through R_v and $\underline{Z}$ becomes minimal. Hereby, the following equations are reduced to a value for R_v = 10 kΩ which gives a cross current of 100 μA at an excitation current of 1 A (0.01%). Looking at this example, the measured impedance $\underline{Z}_m$ is supposed to have a resistance of 300 μΩ and a DC resistance of circa 10 mΩ at high frequencies. Therefore, using the above mentioned values for R_v and R_{shunt} gives:

$$R_s = 300\,\mu\Omega \cdot \frac{R_v + R_{shunt}}{R_{shunt} - 300\,\mu\Omega} \rightarrow R_s = 300\,\mu\Omega \cdot \frac{10\,\text{k}\Omega + 1\,\Omega}{1\,\Omega - 300\,\Omega} \approx 3\,\Omega \tag{2}$$

Similarly, the DC resistance $R_v + R_p$ gives:

$$R_s + R_p = 10\,\text{m}\Omega \cdot \frac{R_v + R_{shunt}}{R_{shunt} - 10\,\text{m}\Omega} \rightarrow R_s + R_p = 10\,\text{m}\Omega \cdot \frac{10\,\text{k}\Omega + 1\,\Omega}{1\,\Omega - 10\,\text{m}\Omega}$$
$$\rightarrow R_p = 101\,\Omega - R_s = 98\,\Omega \tag{3}$$

For finding the parallel capacity C_p, the frequency at the phase minimum must be assumed (i.e. at $\omega = 1$ Hz). For the given impedance $\underline{Z}$ the following expression results:

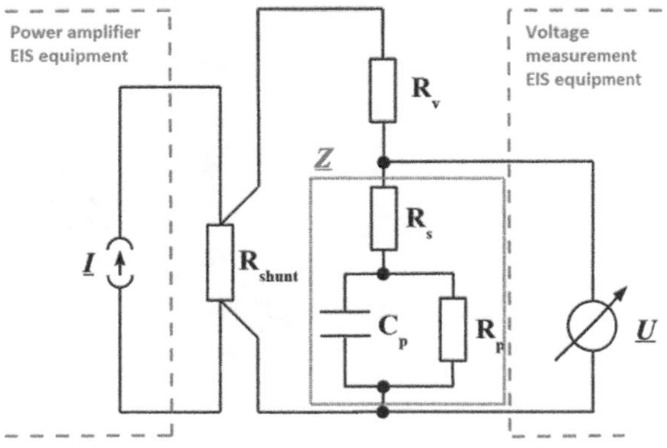

Power amplifier
EIS equipment
Voltage
measurement
EIS equipment
R_v
Z
R_s
I
R_shunt
C_p
R_p
U

Figure 2. Circuit diagram of a reference impedance forming a simplified Randles-element.

$$\underline{Z} = \frac{R_s + R_p + \omega^2 R_p^2 R_s C_p^2}{1 + \omega^2 R_p^2 C_p^2} + j\frac{-\omega R_p^2 C_p}{1 + \omega^2 R_p^2 C_p^2} \tag{4}$$

After the transformation according to (1) the measured impedance $\underline{Z}_m = \frac{U}{I}$ is calculated as:

$$\begin{aligned} \underline{Z}m = & \frac{R_{shunt}(R_s + R_p)(R_s + R_p + R_v + R_{shunt}) + \omega^2 R_p^2 C_p^2 R_s R_{shunt}(R_s + R_v + R_{shunt})}{(R_s + R_p + R_v + R_{shunt})^2 + j\omega^2 R_p^2 C_p^2 (R_s + R_v + R_{shunt})^2} \\ & + j\frac{\omega R_p C_p \left[R_s R_{shunt}(R_s + R_p + R_v + R_{shunt}) - R_{shunt}(R_s + R_p + R_v + R_{shunt})(R_s + R_p) \right]}{(R_s + R_p + R_v + R_{shunt})^2 + \omega^2 R_p^2 C_p^2 (R_s + R_v + R_{shunt})^2} \end{aligned} \tag{5}$$

To find the phase minimum, the equation has to be searched for extremal values regarding the frequency ω:

$$\begin{aligned} & \frac{\mathrm{d}}{\mathrm{d}\omega}\left(\arctan \frac{\Im\{\underline{Z}\}}{\Re\{\underline{Z}\}} \right) \equiv 0 \\ & \frac{\mathrm{d}}{\mathrm{d}\omega}\left(\arctan \frac{-\omega R_p^2 C_p (R_s + R_p + R_v + R_{shunt})}{\omega^2 R_p^2 C_p^2 (R_s + R_v + R_{shunt}) + (R_s + R_p)(R_s + R_p + R_v + R_{shunt})} \right) \equiv 0 \end{aligned} \tag{6}$$

After deriving and converting, C_p results in:

$$\begin{aligned} C_p = & \frac{1}{R_p}\sqrt{\frac{(R_s + R_p)(R_s + R_p + R_v + R_{shunt})}{R_s(R_s + R_v + R_{shunt})\omega^2}} \\ & \rightarrow C_p(\omega = 1\mathrm{Hz}) \approx 1\mathrm{mF} \end{aligned} \tag{7}$$

At 1 Hz there is a phase minimum that corresponds to a parallel capacity of circa 1 mF. After this, all electrical elements are set.

2.3 *Realization*

Figure 3 shows the assembly of the impedance standard. To keep the dimensions of the impedance standard low, only Surface Mounted Devices (SMD) were used. Also, the assembly

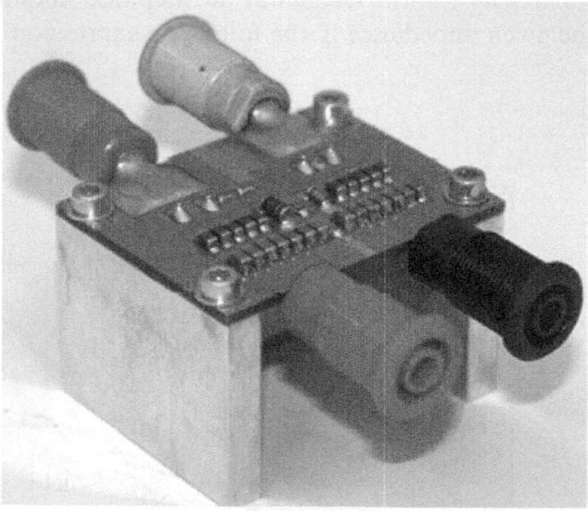

Figure 3. Picture of the assembled circuit. For all passive elements, Surface Mounted Devices (SMD) parts were used, leading to a compact assembly.

gains stability from its flat circuit board structure compared to wire mounted devices. The calculated capacity of C_p of 1 mF is realized by connecting 20 ceramic capacitors in the 1206 SMD package to keep the physical dimension low, as well as the parasitic inductivity.

3 SIMULATION-AND MEASUREMENT RESULTS

The circuit was simulated by LTSpice. The excitation AC current was 1 A and no DC current in the simulation. The simulated frequency range was 10 mHz to 10 kHz. A commercially available impedance meter was used for comparative results, but with the limitation to measure up to 1 kHz in the experiment only.

3.1 *Comparison of measurement and simulation*

The resulting graphs of simulation (dotted blue line) and real measurement (solid red line) are shown in Fig. 4. As can be seen on the Bode and Nyquist plot, the two resulting curves match very well. However, the simulation cannot take any voltage and frequency depending capacity variations into account which are naturally present in ceramic capacitors. Therefore, a small inductance in series was added to influence the simulative results and give a good fit between simulation and measurement.

3.2 *Comparison of various impedance meters*

The described impedance standard shows impedances close to real battery cells for the frequency range of 10 mHz to 10 kHz, which is an expectable range for impedance spectroscopy on batteries. Furthermore, the device acts more predictably regarding temperature and aging behavior than real battery cells, because discrete electronic parts with known and steady attributes were used. Therefore, its behavior is described sufficiently to make it valuable for comparing various EIS measuring equipment. It is intended for using it at various test sites to make sure all measuring devices are calibrated correctly and are operated properly. This procedure had been done on five different impedance meters with the following boundary conditions:

- Frequency range: 1 kHz to 10 mHz
- Spectral density: 8 discrete frequency data point per decade
- 3 periods to be measured per frequency
- At least 1 second of measuring duration per frequency
- Excitation AC current: maximal 1 A
- Voltage response equal or smaller than 10 mV.

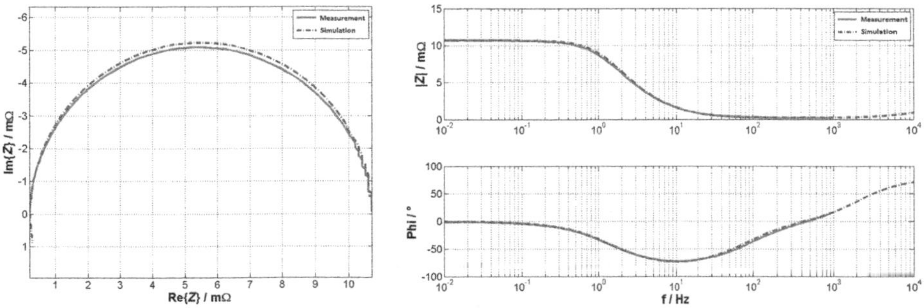

Measurement
Simulation
Im{Z} / mΩ
Re{Z} / mΩ
|Z| / mΩ
Phi / °
f / Hz

Figure 4. Results of simulation and measurement of the reference impedance. On the left side, the results in Nyquist representation are presented, on the right side, the Bode plot is presented.

The five impedance meters used for the comparative test were:

- Zennium Electrochemical Workstation (Zahner Elektronik)
- EISmeter (ISEA)
- EISmeter (Digatron)
- FRA0355 (Maccor)
- VSP300 (Biologic).

The EIS measurement method is based on galvanostatic excitation as long as the 10 mV voltage response is not disturbed. If it is not possible to meet the voltage response limit, a potentiostatic excitation is allowed, too. Furthermore there needs to be a guideline on how to wire the impedance standard. This is important for making sure that the measuring bench does not influence the test [7]. Adding to these guidelines, a proposal to connect individual battery cells in the voltage measuring and current path of the impedance standard is given (see Fig. 5). The voltage visible to the impedance meter's power amplifier is raised in order to strain the power amplifier further.

The coupling between input current and output voltage is taking place within the reference impedance and therefore the added battery cells do not influence the measurement. To make sure there is no damage to the battery cells, it must be ensured to put the cell into a state of charge that overcharging or overdischarging is excluded at any given time.

In Fig. 6 the five devices' and their simulation results are put into the Nyquist plot. Two time constants emerge ($\tau_1 \approx$ 40 ms and $\tau_2 \approx$ 320 ms) and become visible by two superimposed semi circles—opposed to the circuit diagram from Fig. 2, here the same circuit is used, but with two serial R||C elements. At first glance, the measurement results match the simulative approach very well. In Fig. 7 there is a more detailed view of the measurement results. The absolute deviations of magnitude and phase are plotted over the frequency in the graphs top left and top right. Directly below, there are the relative deviations plotted over the frequency.

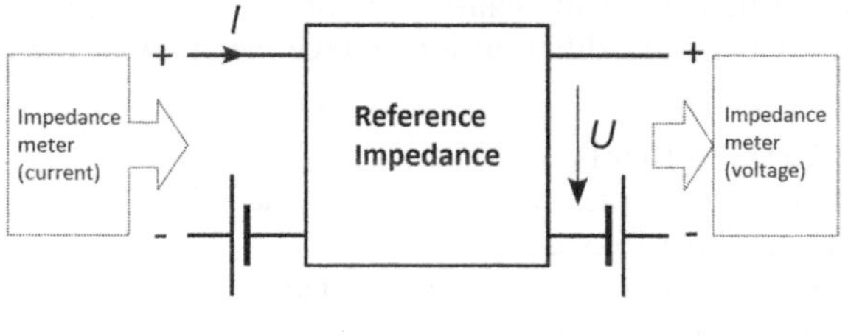
I
+
+
Impedance meter (current)
Reference Impedance
U
Impedance meter (voltage)
-
-

Figure 5. Application of the reference impedance in a comparison experiment. In order to maintain conditions close to reality, a lithium ion cell is connected in series to current input and voltage output.

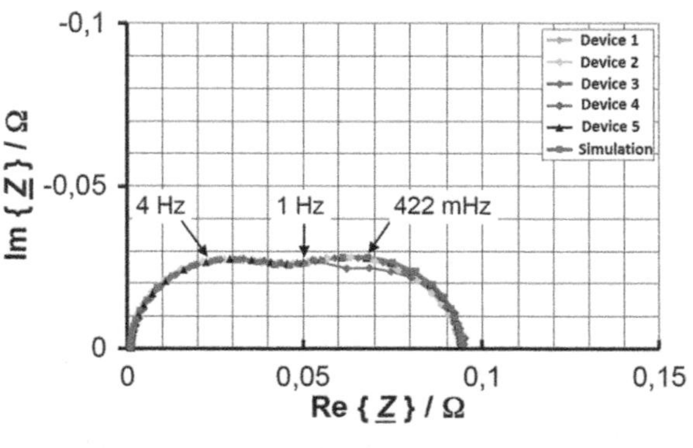
-0,1
-0,05
0
Im {Z} / Ω
0
0,05
0,1
0,15
Re {Z} / Ω
4 Hz
1 Hz
422 mHz
Device 1
Device 2
Device 3
Device 4
Device 5
Simulation

Figure 6. Results of the comparison experiment in Nyquist representation.

The maximal deviation is 6 mΩ (equals 9% at this frequency), whereas the largest relative deviation of the impedance magnitude is at 14%. The reason for the curves' drastically different patterns is the vast acquisition range the impedance magnitude is measured in. On the graph in the bottom left corner of Fig. 7 you can see the simulative impedance/phase plot and the average value curves of the measured results for all measurement equipment. The smallest impedance is 1 mΩ at 1 kHz and the maximal deviation of 14% refers to an absolute value of 140 μΩ.

Very large deviations regarding the phase angle (circa 20°) become dominant in the high frequency range. The absolute deviations tend to decrease towards the lower frequencies (except for one device). The relative deviations are increasing when the measuring frequency decreases, because the absolute value of the phase goes towards 0° in those frequency regions. Hence, the limited resolving capability of individual impedance meters has a stronger impact.

This example points out the difficulty of comparisons of impedance measurements on batteries. At very small impedances, the relative inaccuracy becomes significant, which must be observed closely during dynamical evaluation of impedance based battery models. The phase shows a maximum standard deviation of 11° at the highest frequency (see Figure 7, bottom right). This finding shows the inaccuracy of small impedances at high frequencies when measuring with impedance meters. However, at the same high frequency range, the magnitude has a small standard deviation. At 0.5 Hz the standard deviation is 2.75 mΩ. The impedance standard has a value of 70 mΩ at this frequency. The comparative experiment shows that the impedance meters have a spread of 3.6% on the impedance trace when measuring the reference impedance. A mapping of various equipments is not intended at this point, because the way of functioning is discussed only. Furthermore, the results show the difficulty to reproduce impedance measurements on batteries, even though good care is taken on the test setup. Especially if the measurement results are to be taken to validate impedance based models, the equipment's tolerances must be involved into the accuracy considerations of the models.

The introduced impedance standard is valuable for creating a calibration curve for individual impedance meters to raise the experiment's accuracy.

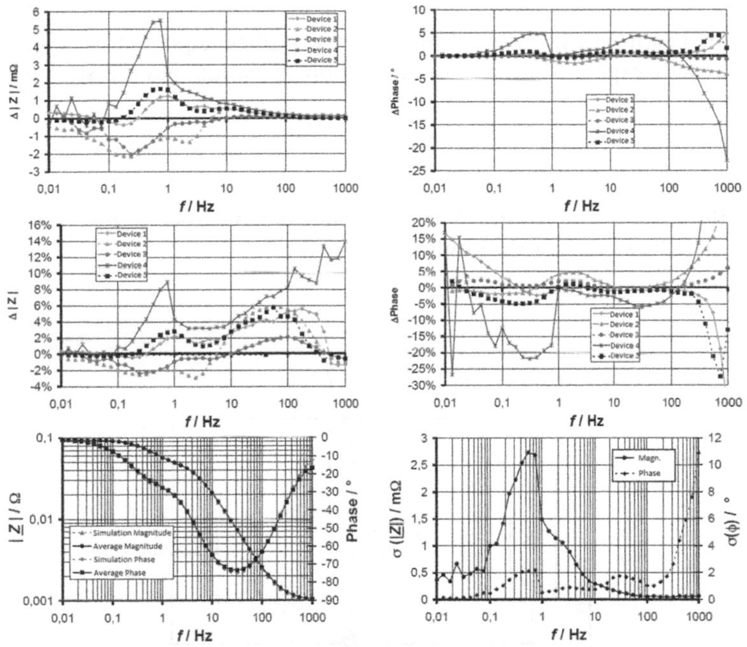

Δ|Z| / mΩ
ΔPhase / °
Δ|Z|
ΔPhase
|Z| / Ω
Phase / °
σ(|Z|) / mΩ
σ(φ) / °
f / Hz
Device 1
Device 2
Device 3
Device 4
Device 5
Simulation Magnitude
Average Magnitude
Simulation Phase
Average Phase
Magn.
Phase

Figure 7. Measurement results of the comparison experiment: top line: Deviation of impedance modulus and phase for each measurement device with respect to the simulation; middle line: Relative deviation for modulus and phase; bottom line: left: Bode plot of the simulation and the average of the 5 measurements; right: Standard deviation for modulus and phase for the 5 different measurement devices.

4 CURRENT DEPENDING NONLINEARITY

In the previous paragraph, the design and the calculation for a reference impedance was given, which reproduces the impedance pattern of an electrochemical cell. However it does not reproduce any dependency regarding current and temperature, which are present in real batteries. By using ceramic capacitors, a parasitic temperature and voltage dependent behavior of the measured impedance is to be expected, but a distinct nonlinear behavior regarding current should be implemented into the impedance standard.

The centerpiece of this design is a nonlinear, current and voltage dependent impedance, which reproduces a charge transfer resistance and therefore must have a response curve close to the Butler-Volmer equation. This is implemented by two diodes connected antiparallel and a resistor connected in parallel. The resistor's task is to adjust the linearity. The subcircuit is shown in Fig. 8. This nonlinear element replaces the resistor R_p (Fig. 8) for the following simulations and results.

The two diodes are of different kinds. *D1* is a generic silicon diode (1N4148), but *D2* is a Schottky type diode (BAT54) which has a lower nominal forward voltage drop than the silicon diode. This differentiation is later used for coping with the different impedance behavior during a charging and discharging regime of a real battery. The equation for the current I is as follows

$$\begin{aligned} I &= I_R - I_{D1} + I_{D2} \\ &= \frac{U}{R} - I_{s1}\left(e^{\frac{-U}{n_1 U_T}} - 1\right) + I_{s2}\left(e^{\frac{-U}{n_2 U_T}} - 1\right) \end{aligned} \tag{8}$$

With I_{s1} and I_{s2} being the reverse currents of $D1$ and $D2$. n_1 and n_2 are the emission coefficients and U_T is the temperature voltage. For very small voltages follows:

$$I = \frac{U}{R} + I_{s1}\left(\frac{I_{s2}}{I_{s1}} e^{\frac{U}{n_2 U_T}} - e^{\frac{-U}{n_1 U_T}}\right) \tag{9}$$

If (9) is compared with the following, simplified Butler-Volmer equation (10), it shows that the system is not exactly explained by the nonlinear diode equation, but has similar properties.

$$I_{ct} = I_0\left(e^{\frac{\alpha\eta}{nU_T}} - e^{\frac{-(1-\alpha)\eta}{nU_T}}\right) \tag{10}$$

By changing the value of the resistor R you can alter the influence of the nonlinearity. This is useful to adjust the impact of the nonlinearity.

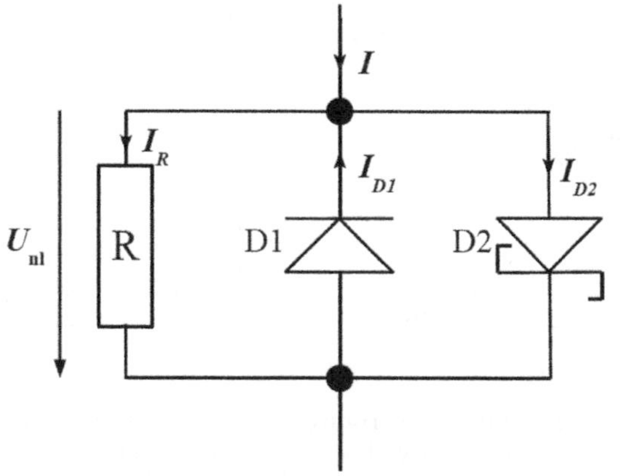
I
I_R
I_D1
I_D2
U_nl
R
D1
D2

Figure 8. Circuit diagram of a nonlinear element.

4.1 *Implementation of the nonlinearity*

For the implementation there have to be a few changes to the already mentioned original circuit. The value of R_v is important for the reference impedance's design—due to the ratio of R_v and the characteristic impedance $\underline{Z}$, there is a voltage drop across both antiparallel diodes. The larger this voltage drop, the stronger gets the nonlinear behavior. This means the voltage across R_v and $\underline{Z}$ must be sufficiently large. This can be done by using a shunt resistor with a larger value. But at the same time, the strain on the power amplifier is increased. To avoid this, a low Ohm shunt resistor in combination with a differential amplifier is used (Fig. 9).

As you can see, the excitation signal is amplified and fed into the voltage divider R_v and $\underline{Z}$. By doing this, the spectra determining network is decoupled from the shunt resistor. This simplifies the calculation and the voltage across the voltage divider and therefore the complex impedance including the nonlinearity circuit can be easily increased. The attached voltage follower and voltage divider is necessary to lower the voltage signal appropriately to get the impedance standard's output impedance into meaningful boundaries. The measured impedance $\underline{Z}_m$ is calculated as follows.

$$\underline{Z}_m = \frac{\underline{U}}{\underline{I}} = \frac{\underline{Z}}{R_v + \underline{Z}} \cdot \frac{R_2}{R_1 + R_2} \cdot R_{shunt} \cdot V_{DiffAmp} \tag{11}$$

As it can be observed, if $R_v \gg Z$, the equation gets simplified further. For example the measured impedance $Z_m = 3.3\ \mathrm{m\Omega}$ for large frequencies, if $R_v = 1\mathrm{k\Omega}$ and $R_s = 3.3\Omega$.

4.2 *Comparison of measurement and simulation*

In Fig. 10 and Fig. 11 a comparison is shown in the Bode and Nyquist plot. LTSpice is the simulation software, where a constant excitation amplitude of 1 A was defined. To show the current dependency, an additional DC current with 2 A, −2 A and 0 A was simulated. Of course, the actual circuit was also measured by an ISEA EISmeter under the same conditions. A measurement with the maximum voltage amplitude (10 mV) was also acquired to complete the experiment.

As you can see, the LTSpice simulation plots the DC current dependency correctly, but the influence of the AC amplitude is distorted. The AC amplitude has a strong influence on the results, which is to be expected according to the theory [8, 9]. The measured curve with 1 A AC amplitude deviates more from the Simulation with 1 A and the curve with 10 mV amplitude is closer to the simulation, for the case without any superposed DC current. This means a better linearity is achieved by continuously adjusting the current amplitude so a maximal voltage response of 10 mV is met. The asymmetric behavior is clearly visible when comparing the curves with the superposed charging and discharging DC currents. In both cases, the measured impedance is smaller than the impedance without any superposed DC current.

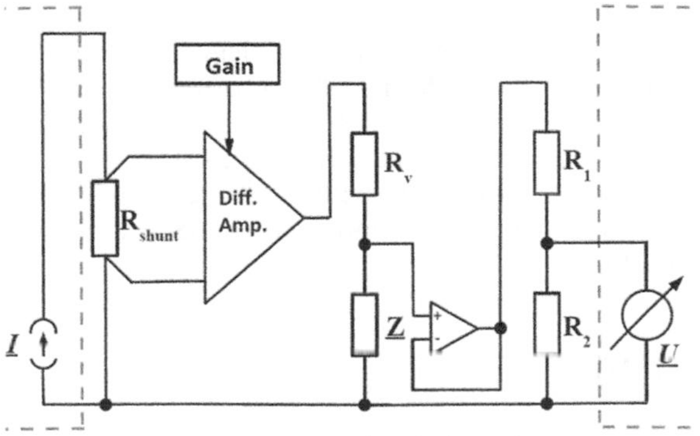

Gain
Diff.
Amp.
R shunt
R v
Z
R 1
R 2
I
U

Figure 9. Schematic circuit diagram of the nonlinear reference impedance.

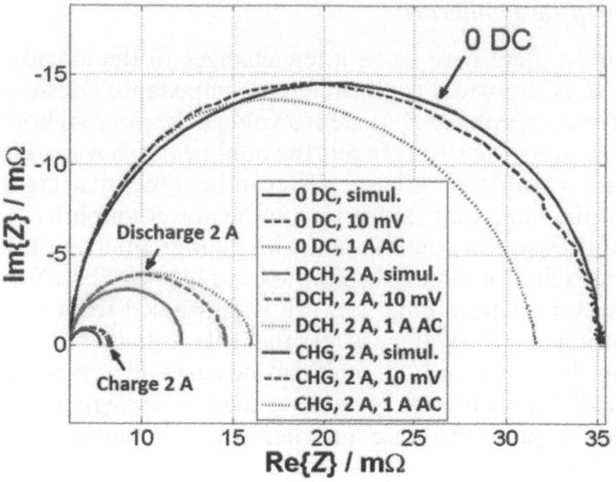
0 DC
Im{Z} / mΩ
-15
-10
-5
0
Discharge 2 A
Charge 2 A
— 0 DC, simul.
--- 0 DC, 10 mV
0 DC, 1 A AC
— DCH, 2 A, simul.
--- DCH, 2 A, 10 mV
DCH, 2 A, 1 A AC
— CHG, 2 A, simul.
--- CHG, 2 A, 10 mV
CHG, 2 A, 1 A AC
10
15
20
25
30
35
Re{Z} / mΩ

Figure 10. Comparitive Nyquist Plot.

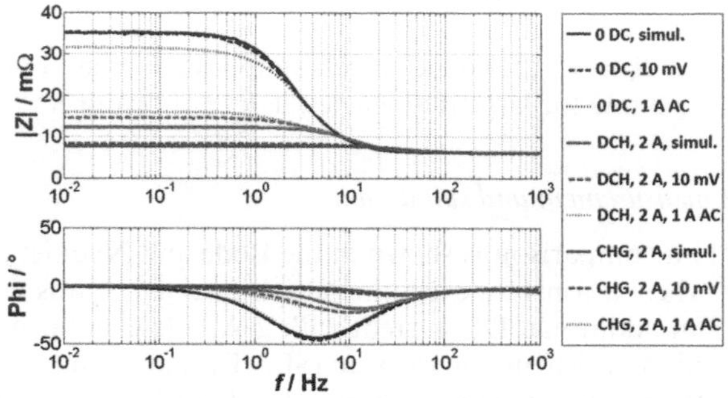
|Z| / mΩ
Phi / °
f / Hz
— 0 DC, simul.
--- 0 DC, 10 mV
0 DC, 1 A AC
— DCH, 2 A, simul.
--- DCH, 2 A, 10 mV
DCH, 2 A, 1 A AC
— CHG, 2 A, simul.
--- CHG, 2 A, 10 mV
CHG, 2 A, 1 A AC

Figure 11. Comparative Bode plot.

5 CONCLUSION

The paper has discussed the need for an impedance standard. At first, the comparison of test results of different impedance measurement devices is mentioned. The design approach is given, which handles the difficulty of large R||C elements in the device's transfer function by the help of a decoupled ouput. The measurement results indicate that for EIS measurements on batteries, it is very difficult to achieve equivalent results from different impedance meters, even if great care is taken on the experiment setup. Especially the phase angle shows a large variance towards high frequencies, which must be taken into account using EIS data for modeling later.

Also, an approach was made, where a nonlinear element was designed by two different diodes to emulate the Butler-Volmer behavior of real batteries, enabling the experimenter to compare different measurement methods on nonlinear test devices. The measurement test result, using the nonlinear element, shows a general consistency with the simulative results. The amplitudes are consistently explainable with the usage of the two different diodes. Further optimizations addressing the electric components values need to be done to get closer to the ideal, simulated results.

ACKNOWLEDGEMENT

The authors thank the European Union for funding the project HELIOS, which brought the opportunity to carry out this work. Also, acknowledgements are directed to the partner institutions involved into this paper's immediate results: *Zentrum für Sonnenenergie—und Wasserstoff-Forschung* (Germany), *Électricité de France* (France), *Ente per le Nuove tecnologie, l'Energia e l'Ambiente* (Italy) and *Austrian Institute of Technology* (Austria).

REFERENCES

[1] J. Gerschler, A. Hammouche, and D.U. Sauer, "Investigation of Cycle-Life of Lithium-Ion Batteries by Means of EIS," *Technische Mitteilungen*, vol. 99, pp. 214–220, 2006.

[2] J. Gerschler, *Ortsaufgelöste Modellbildung von Lithium-Ionen-Systemen unter spezieller Berücksichtigung der Batteriealterung*. PhD thesis, RWTH Aachen University, 2012.

[3] M. Thele, *A Contribution to the Modelling of the Charge Acceptance of Lead-acid Batteries: Using Frequency and Time Domain Based Concepts*. PhD thesis, RWTH Aachen University, 2008.

[4] J. Drillkens, Y. Yurdagel, J. Kowal, and D. Sauer, "Maximizing the lifetime of electrochemical double layer capacitors et given temperature conditions by optimized operating strategies," in *Proceedings of the 4th European Symposium on Supercapacitors and Applications, ESSCAP*, 2010.

[5] H. Walz, "Aufbau und Simulation einer virtuellen Batterie für Impedanzmessungen." Studienarbeit RWTH Aachen University, 2009.

[6] M. Kiel, *Impedanzspektroskopie an Batterien unter besonderer Berücksichtigung von Batteriesensoren für den Feldeinsatz*. PhD thesis, RWTH Aachen, ISEA, 2013.

[7] M. Kiel and D. Sauer, "Impedanzspektroskopie an Batterien," in *Design & Elektronik Entwicklerforum*, 2010.

[8] M. Kiel, O. Bohlen, and D. Sauer, "Harmonic analysis for identification of nonlinearities in impedance spectroscopy," *Electrochimica Acta*, vol. 53, no. 25, pp. 7367–7374, 2008.

[9] K. Darowicki, "The amplitude analysis of impedance spectra," *Electrochimica acta*, vol. 40, no. 4, pp. 439–445, 1995.

Lecture Notes on Impedance Spectroscopy, Volume 4 – Kanoun (ed)

Development of hybrid algorithms for EIS data fitting

Aliaksandr S. Bandarenka
Center for Electrochemical Sciences, Ruhr-Universität Bochum, Bochum, Germany

ABSTRACT: A general approach is proposed for the development of hybrid algorithms which are designed for efficient fitting of EIS data of different origin. The approach aims to construct two-stage hybrid algorithms, in which different minimization strategies are used, reducing both the computational time and the probability to overlook the global optimum. The best candidates to be implemented at each of the stages of the hybrid algorithms are identified by screening for appropriate combinations of optimization strategies. As an application of this approach, in this work, a hybrid iterative algorithm for the analysis of multidimensional EIS data sets has been developed. The developed algorithm is optimized to fit (in a semi-automatic mode) large experimental datasets to equivalent electric circuits commonly used in physical electrochemistry to model interfaces between solid electrodes and liquid electrolytes.

Keywords: EIS data analysis, hybrid algorithms, CNLS, equivalent circuits

1 INTRODUCTION

Classical EIS data analysis consists of two stages: (i) elucidation of an appropriate physical model of the system and (ii) fitting of EIS data to the model (Equivalent Electric Circuit, EEC) to estimate the parameters of the models [1]. The first stage of the data analysis, elucidation of the model, is a difficult task which requires a priori knowledge about the system under investigation as well as long-term expertise in EIS data processing. Nevertheless, long history of EIS enriched this method with efficient theoretical tools for powerful elucidation of physical models of many different classes of electrochemical systems. Moreover, few hundreds of ready-to-use physical models of different electrochemical systems are currently available in the literature. On the other hand, fitting of an impedance spectrum to the elucidated physical model is a well established procedure. However, the latter is valid only for the case when single or several impedance spectra are involved into the data processing.

Routine fitting of impedance spectra becomes considerably more complicated when the amount of EIS data grows significantly [2]. Modern equipment can currently acquire large amount of impedance spectra fast enough to collect EIS data as a function of time or electrode potentials at rates suitable to investigate markedly non-stationary electrochemical processes. One example of such a dataset, which consists of more than 1000 spectra, is shown in Figure 1 (the dataset was acquired during Cu intercalation into Te electrode). However, fitting of EIS spectra to the electrochemical models remains nowadays largely time consuming, where each single spectrum is processed individually. Practically, this means that due to complexities in the fitting of large EIS datasets the opportunities which are provided by modern equipment are not fully used. Therefore, one of the main impediments to a wider and more efficient application of EIS-based approaches in different areas is the ability to process large data sets of impedance data on a reasonable timescale.

Commercially available software developed to process individual impedance spectra use few general algorithms such as Levenberg-Marquardt algorithm, the Nelder-Mead downhill simplex method or genetic algorithms [3–7]. The software is optimized to process only

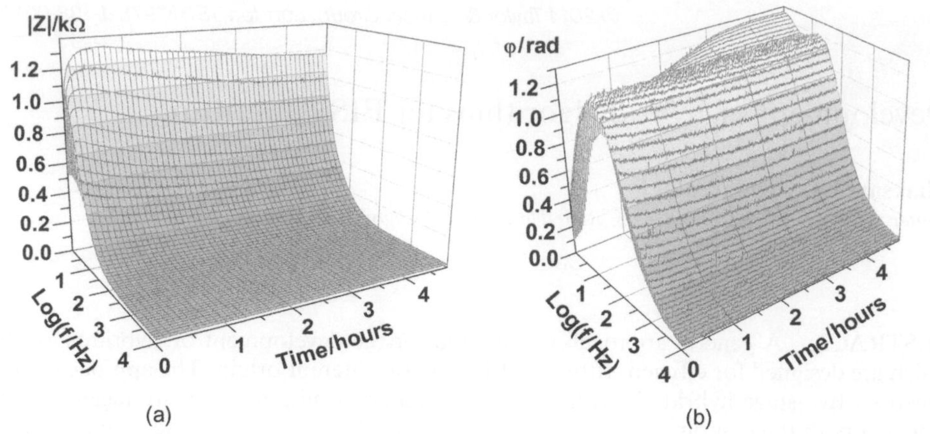
|Z|/kΩ
1.2
1.0
0.8
0.6
0.4
0.2
0.0
Log(f/Hz)
Time/hours
(a)
φ/rad
1.2
1.0
0.8
0.6
0.4
0.2
0.0
Log(f/Hz)
Time/hours
(b)

Figure 1. An example of a large EIS experimental dataset, which is a collection of impedance spectra acquired during Cu intercalation into tellurium electrode. The dataset consists of 1005 spectra, where each singe spectrum consists of 50 frequency points. (a) The dependence of the modulus of impedance and (b) the dependence of the phase shift as a function of probing frequency and time.

individual spectra. Moreover, the fitting algorithms are not normally optimized to process EIS data of different origin.

In this work, a general approach is proposed for the development of hybrid algorithms designed for efficient fitting of (large) experimental EIS datasets of different origin. As an application of this approach, a hybrid algorithm for the fitting of large EIS data sets has been constructed and optimized.

2 DEVELOPMENT OF HYBRID ALGORITHMS

The proposed approach aims to develop a two-stage hybrid algorithm providing stable sequential fitting or EIS spectra. The idea is based on a following hypothesis: if the most powerful existing optimisation strategies cannot provide the required stability, universality and reasonable computational time for the analysis of large experimental EIS datasets, it might be possible to construct and optimize two-stage hybrid algorithms (see Figure 2). At Stage I of the hybrid algorithm, it is important to have an optimization strategy capable of efficiently identifying the general area of the global minimum based on *a priori* initial values obtained at some previous stages or based on predefined by the user upper and lower parameter values (taking into account physical meaning of the parameters). This Stage I must always provide parameter values which are reasonably close to the global minimum irrespective of function topology, influence of random noise and any other relevant factors. At Stage II, the algorithm should be able to identify the exact position of the minimum.

To construct an efficient two-stage hybrid algorithm, a screening procedure should be performed involving (for the simplicity and as a first approximation) some known optimisation algorithms and strategies. One example of such a screening is given below aiming to construct the hybrid algorithm capable to process large experimental datasets.

Table 1 shows the output of the screening procedure, where 6 common and widely used algorithms were tested in different combinations (36 combinations in total). To find the optimal combinations, different types of algorithms and their combinations were selected. These are: (i) optimisation (Simplex), iterative (Levenberg-Marquardt [8, 9]), and (ii) heuristic (Nelder-Mead [10]) methods and (1) the non-gradient (Nelder-Mead, Simplex, Genetic [11]) and (2) gradient (Levenberg-Marquardt, BFGS) methods.

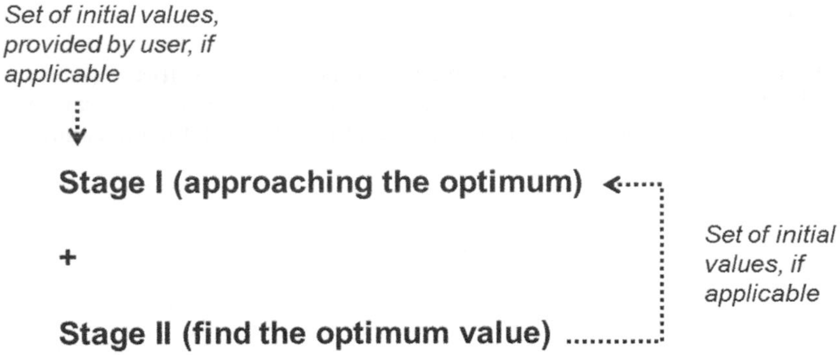
Set of initial values, provided by user, if applicable
Stage I (approaching the optimum)
+
Set of initial values, if applicable
Stage II (find the optimum value)

Figure 2. Schematics illustrating the general structure of hybrid algorithms for (semi) automatic fitting of large experimental EIS datasets.

Table 1. Output of the algorithm screening procedure.

	Stage II					
Stage I	Simplex	Levenberg-Marquardt	Nelder-Mead	BFGS	Powell	Genetic algorithms
Simplex	x	x	x	x	x	x
Levenberg-Marquard	x	x	s	x	x	x
Nelder-Mead	x	s		x	x	x
BFGS	x	x		x	x	x
Powell	x	s		x	x	x
Genetic algorithms	x	x		x	x	x

The evaluation criteria can be different and depend on specific tasks and data origin. In this work, the evaluation criteria were selected as follows: (i) the computational time, (ii) accurate identification of the global minima of the functions of different topologies, and (iii) stability for the fitting of large experimental EIS data-sets. If (i) the total computational time was $\leq$ 20s per spectrum (3 GHz processor, 4GB RAM), (ii) the algorithm combination was able to work accurately with functions of different topologies (practically, different EECs were used) and (iii) it fails in finding optima less than one time per 20 spectra in the automatic fitting mode using initial values from the previously identified minima, then the algorithm combination was evaluated as "good" (marked with grey in Table 1). If the algorithm combination does not pass the criterion (ii), it was evaluated as "satisfactory" (s). Other cases were evaluated as "not acceptable" (x).

As can be seen from Table 1, two promising combinations were found. These two algorithm combinations were tested further using simulated impedance data-sets to construct and optimize the hybrid algorithm. The "Nelder-Mead + Nelder-Mead" combination is obviously a trivial combination, which cannot be optimised further in the framework of the developed approach. This algorithm was used to process data in potentiodynamic electrochemical impedance spectroscopy [12]; however it has a very limited applicability to process larger datasets, especially when complex models (equivalent electric circuits) are used. Therefore, only one combination ("Powell + Nelder-Mead" in Table 1) was further optimized as described in detail elsewhere [2]. Briefly, Powell's optimisation strategy was modified in order to increase the probability to localise the area of the global minima by additionally exploring wide areas of the parameter space during each single iteration. The second stage, which involves the Nelder-Mead strategy, was restricted to the area of the optimum identified during the first stage (± 3% from the optimum found at the first stage).

3 IMPLEMENTATION

The resultant hybrid algorithm was implemented in home-made software (see screenshots in Figure 3). The software consists of 3 logical blocks in which the user can explore the raw data in different representations (imaginary and real parts of the impedance, absolute

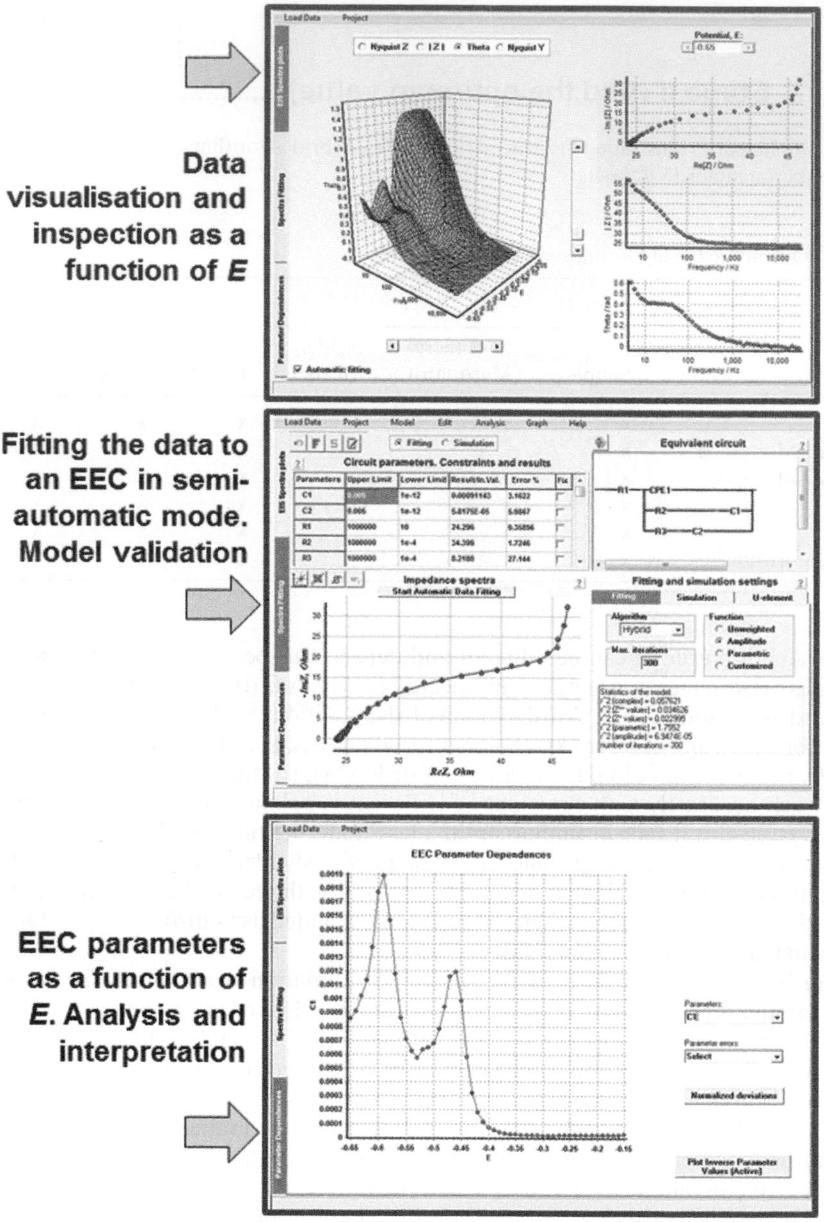

Data visualisation and inspection as a function of *E*
Fitting the data to an EEC in semi-automatic mode. Model validation
EEC parameters as a function of *E*. Analysis and interpretation
Circuit parameters. Constraints and results
Equivalent circuit
Impedance spectra
Fitting and simulation settings
EEC Parameter Dependences

Figure 3. Screenshots of software for large experimental EIS-datasets processing which uses the hybrid optimization algorithm developed in this work. The experimental dataset (impedance spectra as a function of the electrode potential) was acquired using Pt electrode in sulfuric acidic medium during the anodic potential scan. See [13] for further experimental details.

values of the impedance, and the phase shift as a function of the electrode potential, time spatial coordinates etc., depending on the origin of datasets), fit the data in an automatic mode using the developed algorithm, and explore the resultant dependences of the EEC-parameters (Figure 3).

4 EVALUATION

The following data-set was simulated using the equivalent electric circuit shown in Figure 4 (a) with EIS Spectrum Analyser software [12]: *Re*(*Z*); *Im*(*Z*); probing AC frequency; *x*-coordinate; *y*-coordinate.

The simulated dataset can be considered as a collection of 121 individual impedance spectra acquired during localised electrochemical impedance spectroscopy measurements at the surface of a hypothetical working electrode during scans along *x* and *y* directions. 25 frequencies (60 kHz–10 Hz) were used to simulate the EIS data.

Figures 4 (b)–(d) show typical individual impedance spectra from the simulated dataset together with corresponding fitting results. Figure 5 (a) schematically shows the hypothetical

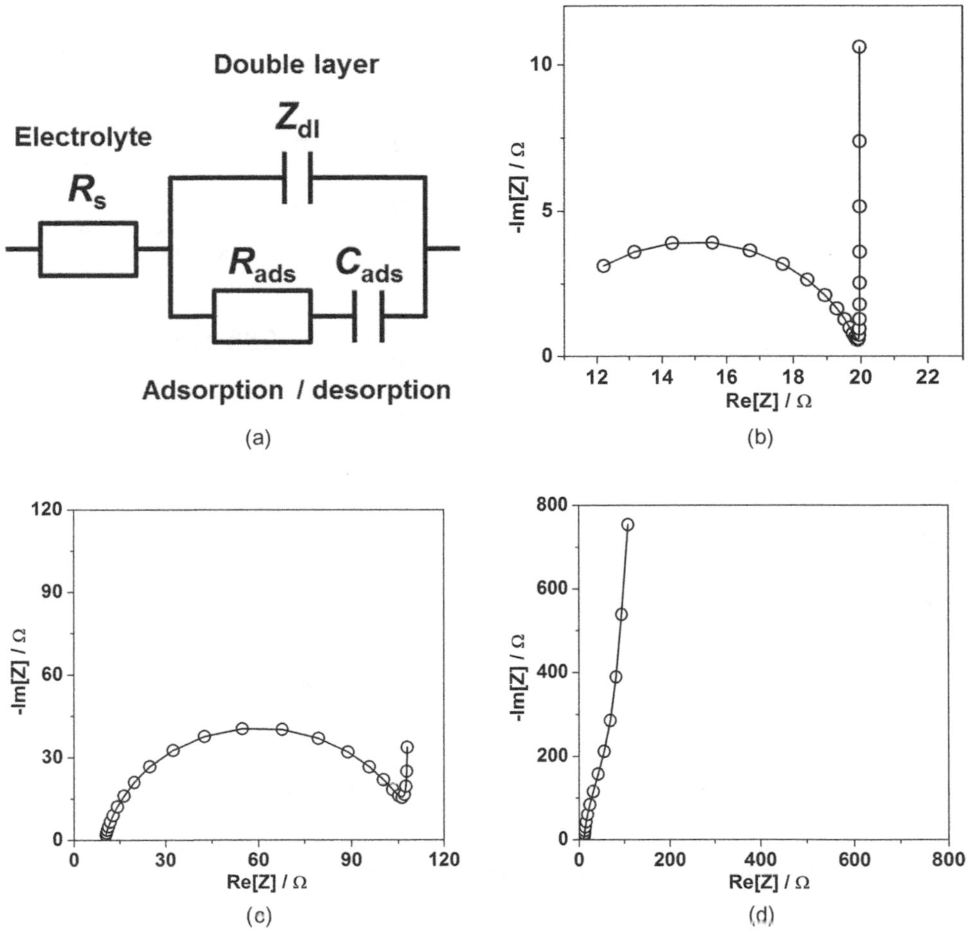

Double layer
Z_{dl}
Electrolyte
R_s
R_{ads}
C_{ads}
Adsorption / desorption
(a)
-Im[Z] / Ω
Re[Z] / Ω
(b)
-Im[Z] / Ω
Re[Z] / Ω
(c)
-Im[Z] / Ω
Re[Z] / Ω
(d)

Figure 4. (a) The equivalent electric circuit used for simulation of large EIS datasets and (b)–(d) examples of simulated individual impedance spectra with fitting results.

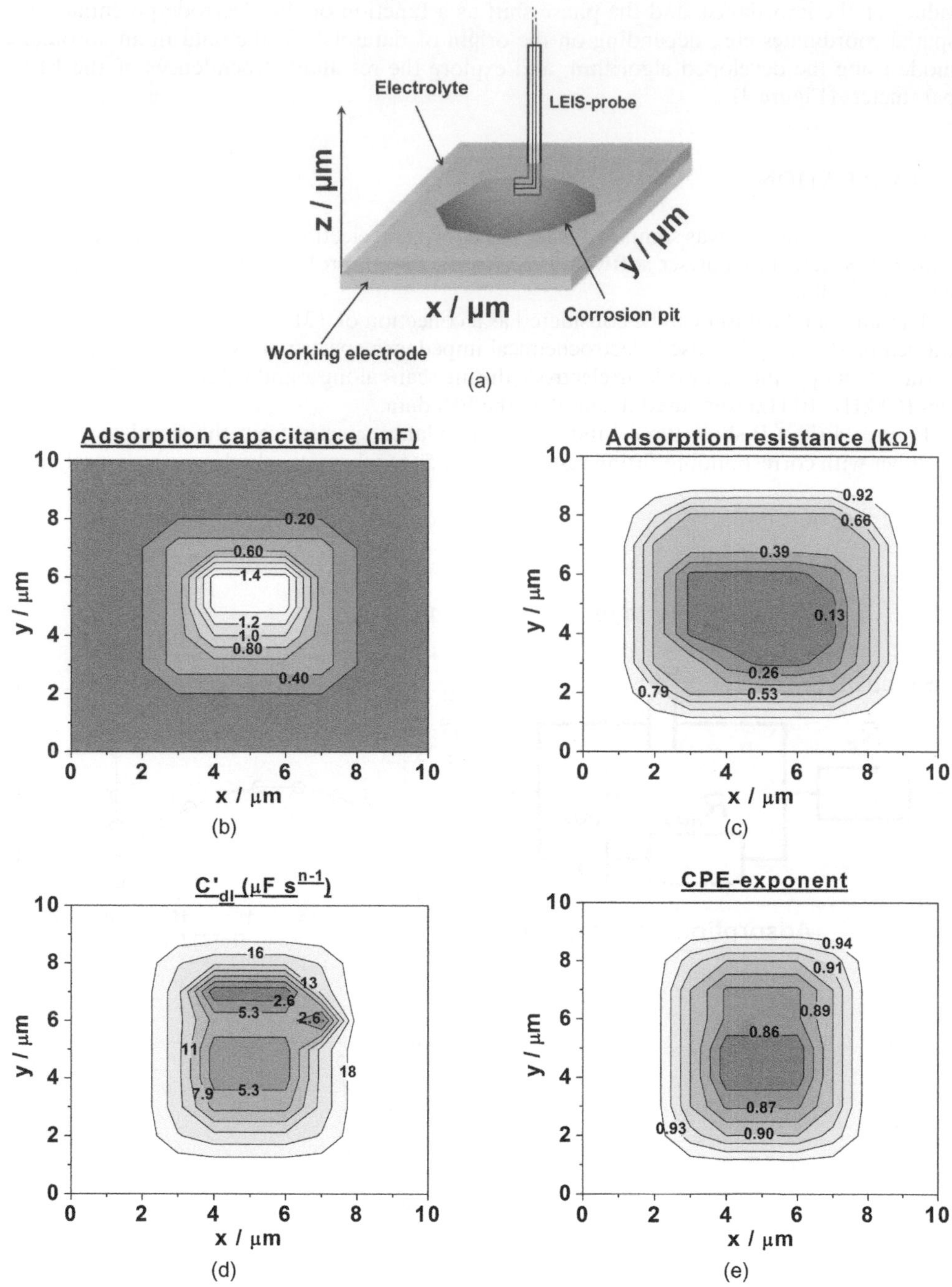
Electrolyte
LEIS-probe
z / µm
y / µm
x / µm
Corrosion pit
Working electrode
(a)
Adsorption capacitance (mF)
Adsorption resistance (kΩ)
C'dl (µF s^n-1)
CPE-exponent
x / µm
y / µm
(b)
(c)
(d)
(e)

Figure 5. (a) A hypothetical electrode surface and (b)–(e) the EEC-parameter dependences shown as a function of spatial coordinates x and y (121 data-points). The parameter values were extracted from the simulated data-set using the approach developed in this work. The values found by the developed algorithm were exactly the same as used for the EIS data simulation.

electrode surface at which a corrosion process takes place and which has been used for the simulations of the large EIS dataset. Figures 5 (b)–(e) present the dependences of the EEC (Figure 4 (a)) as a function of spatial coordinates, x and y, reconstructed using the developed hybrid algorithm. The algorithm demonstrated good stability of fitting in automatic mode. The calculated values of the model parameters were exactly those same as used for the EIS data simulation. The estimated errors of the parameter values were less than 0.0001%. The computational time to process the simulated dataset was 3 min 30s using a 3 GHz processor (4 GB RAM). The computation was performed in a background mode so that the user could use the computer for other tasks. The algorithm was also successfully tested using a larger simulated dataset (300 impedance spectra, 50 frequency points per spectrum) simulated with Randles equivalent circuit [2].

Additionally, the developed hybrid algorithm was evaluated using different experimental datasets describing (i) deposition of atomic layers of metals in presence of specific adsorption of other electrolyte components [14], (ii) an electrocatalytic reaction (nitrate reduction) which occur simultaneously with formation of the catalyst layers [15], (iii) where adsorption and desorption of different anions and cations occur simultaneously [13], (iv) during formation and growth of thin metal films [16]. The algorithm showed excellent performance in processing of these large experimental datasets.

It should be noted, however, that the ability of the developed hybrid algorithm to process large experimental datasets in semi-automatic mode within reasonable timescale is defined by the selection criteria (i–iii) as stated in Section 2 and the datasets. The selection criteria can be modified according to the requirements or specific tasks. For instance, if computational time is not a critical issue, genetic algorithms can be also used at the Stage I. This can further improve stability of the hybrid algorithms as genetic algorithms normally determine the area of the global minimum with higher probability compared to other known optimization strategies.

5 CONCLUSION

An approach is proposed for the development and optimisation of hybrid fitting algorithms for efficient processing of impedance data. In a simple case, a two-stage hybrid algorithm can be constructed in which different minimization strategies are used, reducing both the computational time and the probability to overlook the global optimum. The minimisation strategies to be implemented at each of the stages of the hybrid algorithms should be identified by screening for the appropriate combinations of known optimisation algorithms. As an application of the developed approach, a hybrid iterative algorithm for the analysis of multidimensional EIS data sets has been developed. The developed algorithm is optimized to process large experimental datasets to equivalent electric circuits commonly used in physical electrochemistry. The algorithm demonstrated excellent performance in fitting both simulated and experimental datasets and can be further modified for other impedance applications which require processing of large experimental datasets.

REFERENCES

[1] M.E. Orazem and B. Tribollet, *Electrochemical Impedance Spectroscopy*, vol. 48. Wiley-Interscience, 2011.

[2] A.S. Bondarenko, "Analysis of Large Experimental Datasets in Electrochemical Impedance Spectroscopy," *Analytica Chimica Acta*, vol. 743, pp. 41–50, 2012.

[3] B.A. Boukamp, "A nonlinear least squares fit procedure for analysis of im-mittance data of electrochemical systems," *Solid State Ionics*, vol. 20, no. 1, pp. 31–44, 1986.

[4] B.A. Boukamp, "A package for impedance/admittance data analysis," *Solid State Ionics*, vol. 18, pp. 136–140, 1986.

[5] J.R. Macdonald and L.D. Potter, "A flexible procedure for analyzing impedance spectroscopy results: Description and illustrations," *Solid State Ionics*, vol. 24, no. 1, pp. 61–79, 1987.

[6] J.R. Macdonald, "Comparison and application of two methods for the least squares analysis of immittance data," *Solid State Ionics*, vol. 58, no. 1, pp. 97–107, 1992.

[7] *ZSimpWin, Version 3.30*, 2004.

[8] D.W. Marquardt, "An algorithm for least-squares estimation of nonlinear parameters," *Journal of the Society for Industrial & Applied Mathematics*, vol. 11, no. 2, pp. 431–441, 1963.
[9] W.H. Press, S.A. Teukolsky, W.T. Vetterling, and B.P. Flannery, *Numerical Recipes in C++: The Art of Scientific Computing*. Cambridge University Press, 2002.
[10] J.A. Nelder and R. Mead, "A simplex method for function minimization," *The computer journal*, vol. 7, no. 4, pp. 308–313, 1965.
[11] T. VanderNoot and I. Abrahams, "The use of genetic algorithms in the nonlinear regression of immittance data," *Journal of Electroanalytical Chemistry*, vol. 448, no. 1, pp. 17–23, 1998.
[12] A.S. Bondarenko and G.A. Ragoisha, *Progress in Chemometrics Research*, ch. 7, pp. 89–102. Nova Science Publishers, 2005.
[13] B.B. Berkes, G. Inzelt, W. Schuhmann, and A.S. Bondarenko, "Influence of Cs^+ and Na^+ on Specific Adsorption of $^{\bullet}OH$, $^{\bullet}O$, and $^{\bullet}H$ at Platinum in Acidic Sulfuric Media," *The Journal of Physical Chemistry C*, vol. 116, no. 20, pp. 10995–11003, 2012.
[14] B.B. Berkes, A. Maljusch, W. Schuhmann, and A.S. Bondarenko, "Simultaneous Acquisition of Impedance and Gravimetric Data in a Cyclic Potential Scan for the Characterization of Nonstationary Electrode/Electrolyte Interfaces," *The Journal of Physical Chemistry C*, vol. 115, no. 18, pp. 9122–9130, 2011.
[15] M. Huang, J.B. Henry, B.B. Berkes, A. Maljusch, W. Schuhmann, and A.S. Bondarenko, "Towards a detailed in situ characterization of non-stationary electrocatalytic systems," *Analyst*, vol. 137, no. 3, pp. 631–640, 2012.
[16] B.B. Berkes, J.B. Henry, M. Huang, and A.S. Bondarenko, "Electrochemical characterisation of copper thin-film formation on polycrystalline platinum," *ChemPhysChem*, vol. 13, no. 13, pp. 3210–3217, 2012.

Bioimpedance modeling and characterization

Lecture Notes on Impedance Spectroscopy, Volume 4 – Kanoun (ed)
© 2014 Taylor & Francis Group, London, ISBN 978-1-138-00140-4

Evaluation of a swine model for impedance cardiography—a feasibility study

M. Ulbrich
Philips Chair for Medical Information Technology (MedIT), RWTH Aachen University, Aachen, Germany

J. Mühlsteff
Philips Research, Eindhoven, The Netherlands

S. Weyer, M. Walter & S. Leonhardt
Philips Chair for Medical Information Technology (MedIT), RWTH Aachen University, Aachen, Germany

ABSTRACT: Impedance Cardiography (ICG) is a simple and cheap method to acquire hemodynamic parameters. In this work, the feasibility of a swine model to extract these parameters with regard to reflect human impedance measurements is analyzed. Therefore, not only the physiological morphology of the signal shall be analyzed, but also its behavior during fluid accumulation. For this purpose, the bioimpedance of female pigs have been monitored for 5:30 h while being infused for blood volume stabilization. Our work uses data collected at experiments dedicated on research on optimal extracorporal circulation while ICG was used as an additional measurement value for cardiovascular supervision. The results show a linear correlation between the calculated total body water and the measured fluid balance. In addition, the morphology of the ICG signals is comparable to human signals, providing the possbility to extract stroke volumes using modified model assumptions. Furthermore, cardiac output measured invasively shows the same trend as estimated by bioimpedance.

Keywords: Impedance Cardiography, bioimpedance, swine model, ECMO, fluid balance, animal study, cardiac output, stroke volume

1 INTRODUCTION

Cardiovascular diseases are the most common cause of death in Western Europe. Possible measures of severity are hemodynamic parameters such as Stroke Volume (SV) or Cardiac Output (CO) which quantify the ability of the heart to supply the body with blood.The gold standards to assess these parameters are, on the one hand, an invasive technique using catheters, and, on the other hand, a non-invasive technique using ultrasound which needs specialists for interpretation. As an alternative, another non-invasive and cheap technnique which can be used by non-trained personnel is ICG. However, this technology is not commonly used in clinical practice, since it is not considered to provide reliable results [1]. One reason for that is the inaccuracy of the technology measuring patients with cardiovascular diseases concerning absolute SV calculations. A swine model could provide the opportunity to analyze processes inside the body occuring along with certain diseases which influence the measured impedance signal.

In the following, the possibility to use a swine model reflecting human impedance measurements will be analyzed.

2 BIOIMPEDANCE MEASUREMENTS

ICG is a bioimpedance measurement method which assesses the impedance continuously at one frequency, usually around 100 kHz. In conventional bioimpedance measurement techniques which try to assess the overall body composition, the frequency spectrum between 5 kHz and 1 MHz (called β-dispersion range) is used, since pathological and pathophysiological effects, like fluid shifts, have the biggest influence on the measured signal in this range. When measuring biological tissues spectroscopically, the complex impedance curve (frequency locus plot) is a depressed semi-circle called Cole plot [2]. Since ICG operates at a fixed frequency, only one continuous point on a complex Cole curve is obtained by ICG measurements (see fig. 1).

Hence, ICG measurements look at the signal pattern in time, not so much on the spectroscopic properties. In practice, two outer electrodes inject a small alternating current into the thorax and by two inner electrodes, the voltage drop is measured. In impedance cardiography, only the magnitude of the complex impedance is analyzed since real and imaginary part are supposed to contain the same information. Using the temporal derivative of the signal $\left|\frac{dZ}{dt}\right|$, time-dependent hemodynamic parameters can be extracted from the measured impedance curve (see fig. 2).

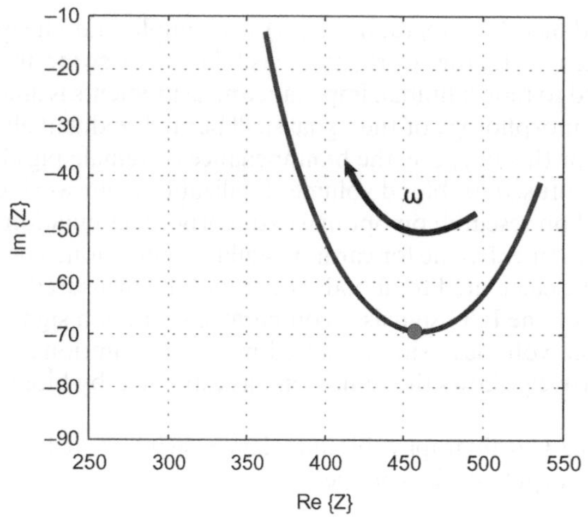

−10
−20
−30
−40
−50
−60
−70
−80
−90
Im {Z}
ω
250
300
350
400
450
500
550
Re {Z}

Figure 1. Cole plot with marked ICG frequency.

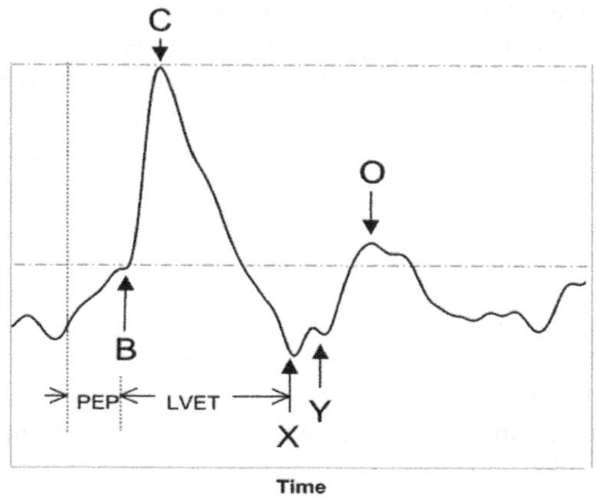

C
O
dZ/dt
B
PEP
LVET
X
Y
Time

Figure 2. ICG wave $(-z(t))'$ with characteristic points [3].

Its maximum is used for the calculation of SV. The SV measured by ICG according to Bernstein and Sramek can be described by the following equation [4]:

$$SV = \delta \cdot \frac{(0{,}17)^3}{4{,}2} \cdot \left|\frac{dZ}{dt}\right|_{max} \cdot \frac{LVET}{Z_0} \tag{1}$$

Here, the factor δ is the actual weight divided by the ideal weight, t_e the Left Ventricular Ejection Time (LVET) and Z_0 the thoracic base impedance. CO is calculated by multiplying SV and Heart Rate (HR): CO = SV · HR.

3 METHODS

In the presented study, female farm pigs (approx. 58 kg, 100 cm) were anaesthesized and connected to a veno-venous extracorporal circulation circuit. They were intubated and ventilated in volume controlled mechanical ventilation mode. In order to maintain hemodynamic stability, a continuous infusion of physiologic saline solution was applied. Every 15 minutes, the amount of supplied infusion was documented using the liquid level scale on the infusion bottle. Thermal management with blanket heating and heating pads ensured constant body temperature. Standard Intensive Care Unit (ICU) instrumentation was installed (ECG, Arterial Blood Pressure (ABP), Venous Blood Pressure (VBP), partial pressure of oxygen (SpO_2), temperature) and continuous CO and HR were measured with a Baxter Vigilance monitoring system using a pulmonary artery catheter.

In addition, the bladder was catheterized to monitor the fluid balance in combination with the amount of the infused saline solution. During the whole study, ICG was assessed using standard adhesive electrodes and standard electrode positions (see fig. 3). The measurement device was the Niccomo device, manufactured by medis GmbH, Ilmenau, Germany, which injects a sinusoidal current of 1.5 mA at 85 kHz. In total, the study lasted 5 hours and 30 minutes.

4 RESULTS

To evaluate the signal quality of the swine model, the morphology of the signal had to be analyzed. Therefore, the raw signal was filtered with a bandpass filter (0.8 Hz < f < 10 Hz). Since the ventilation caused a non-sinusoidal modulation of the impedance signal, only one heartbeat after inspiration has been analyzed. An example is shown in fig. 4.

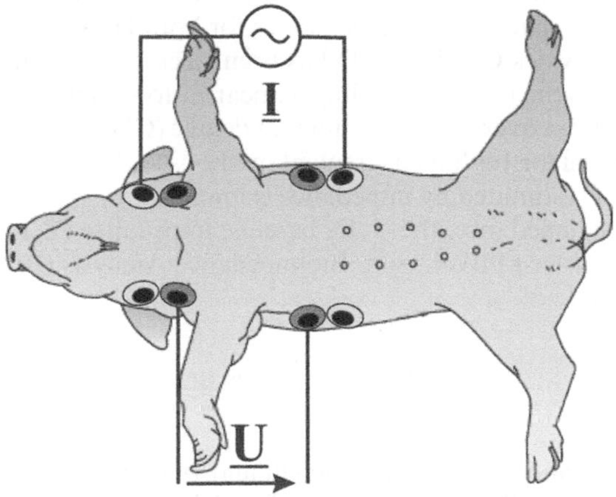

I
U

Figure 3. Pig with electrode positions.

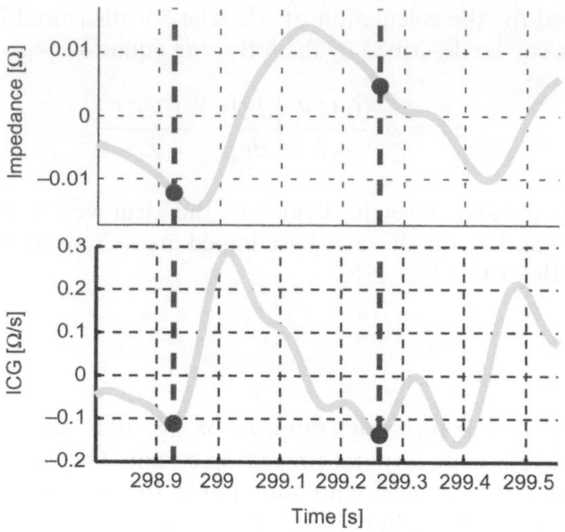

Impedance [Ω]
0.01
0
−0.01
ICG [Ω/s]
0.3
0.2
0.1
0
−0.1
−0.2
298.9
299
299.1
299.2
299.3
299.4
299.5
Time [s]

Figure 4. Measured impedance and ICG with B- and X-point.

The graph shows that the measured impedance as well as its derivative is 10 times lower than in humans (peak-to-peak 0.03 Ω). Nevertheless, all important characteristic points can be extracted. In fig. 4 one can see the successful extraction of the standard B- and X-point, resulting in an LVET of 335 ms which is in a normal range. Although there are other possibilities to extract characteristic points, only one approach has been used for this work. The B-point was assessed using the local minimum before the C-point and the X-point using the global minimum after the C-point in a certain time interval [5, 6].

The base impedance, which is not visualized in fig. 4, ($17.3\ \Omega < Z_0 < 19.6\ \Omega$) is approximately two times lower compared to measurements on the human thorax. Since the SV equation is proportional to $\left|\frac{dZ}{dt}\right|_{max}$ and antiproportional to Z_0 (see eq. 1), an underestimation of the real SV is expected.

When analyzing the whole measurement period, the measured HR by invasive ECG and ICG visually correlate perfectly but show a temporal shift (see fig. 5a).

Non-synchronized measurement devices are a common problem but one would expect a linear correlation between time stamps of two systems with a slightly different system clock. But in this case, a highly non-linear temporal shift was observed (see fig. 6).

After the identification of the temporal compensation factors (m-factors), the time stamps for one system were compensated in order to be able to correlate all measured values over the whole measurement time. The compensated curves for both HRs are shown in fig. 5b. Since the Vigilance system provides CO data, CO has been calculated by impendance to compare both values. Fig. 7 shows the results including a linear fit to visualize the global trend.

Both CO curves decrease over time by a different degree ($CO_{Vigilance}$: −53%, $CO_{Niccomo}$: −30%) and since the HR is equal for both measurement methods and SV is expected to be underestimated, we see that CO estimated by impedance is lower.

Besides the dynamic impedance, the static baseline impedance (Z_0) has been analyzed. In literature, Total Body Water (TBW) using Bioimpedance Analysis (BIA) is calculated using the following approach:

$$TBW = a \cdot \frac{L^2}{Z_0} + b \cdot W \tag{2}$$

Here a and b are constants, L is the body height and W body weight [7]. For our purposes, a has been set to 0.2 and b to 0.25. The calculated TBW was then compared to the measured fluid balance. The results are shown in fig. 8(a).

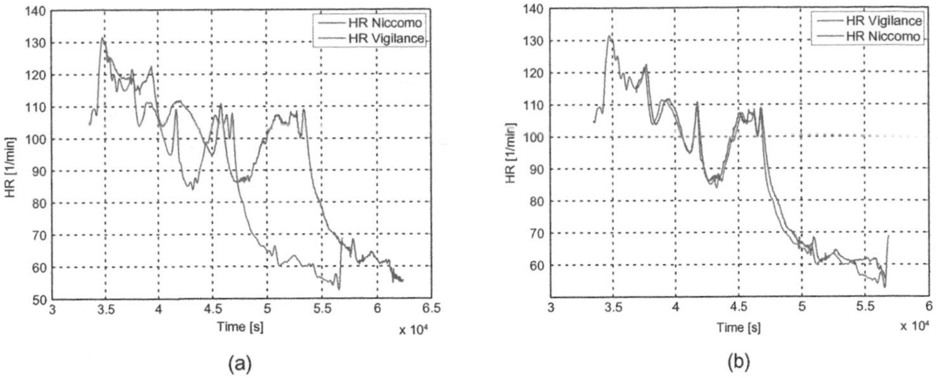
HR Niccomo
HR Vigilance
HR [1/min]
Time [s]
x 10⁴
(a)
HR Vigilance
HR Niccomo
HR [1/min]
Time [s]
x 10⁴
(b)

Figure 5. Heart rates: (a) uncompensated; (b) compensated.

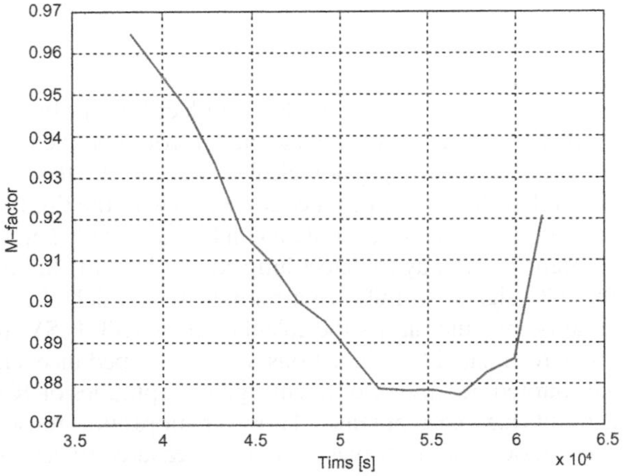
M-factor
Tims [s]
x 10⁴

Figure 6. Temporal compensation factors.

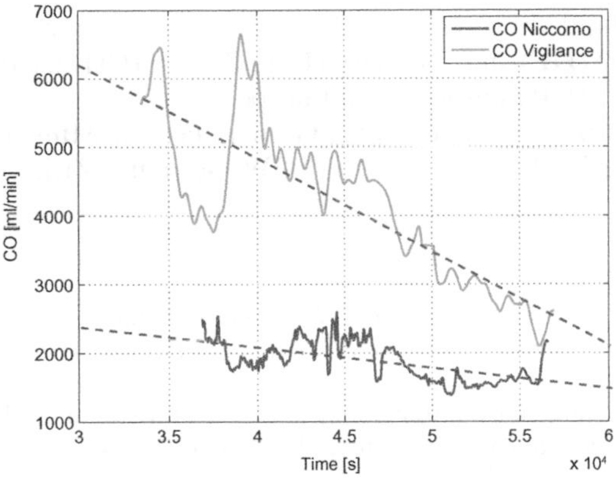
CO Niccomo
CO Vigilance
CO [ml/min]
Time [s]
x 10⁴

Figure 7. Cardiac outputs.

This graph shows that the measured fluid balance is strongly correlated with the calculated body water ($r^2 = 0.9324$). When taking this correlation into account, both measurement methods have been compared using a Bland-Altman plot (see fig. 8 (b)). It shows that the mean error is low and all measured values lie within the limits of agreement ($1.96 \cdot \sigma$), where σ is the standard deviation of the mean value μ.

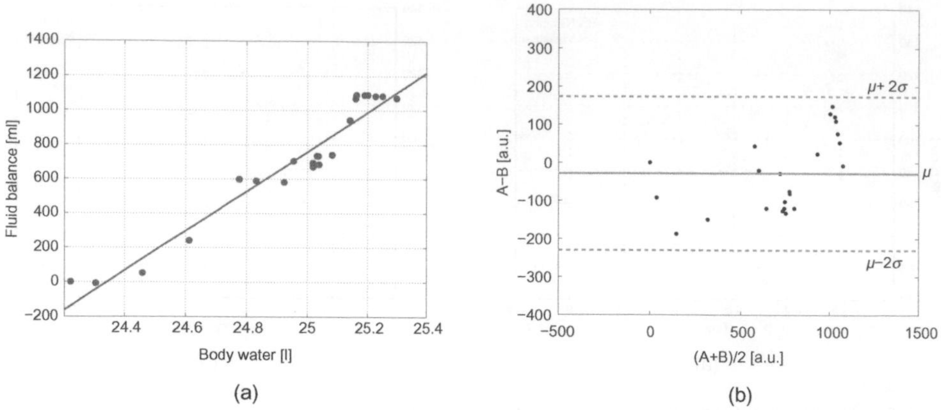
1400
1200
1000
800
600
400
200
0
−200
Fluid balance [ml]
24.4
24.6
24.8
25
25.2
25.4
Body water [l]
(a)
400
300
200
100
0
−100
−200
−300
−400
A−B [a.u.]
μ+ 2σ
μ
μ−2σ
−500
0
500
1000
1500
(A+B)/2 [a.u.]
(b)

Figure 8. (a) Measured fluid balance vs. calculated body water; (b) Bland-Altman plot for fluid balance.

5 CONCLUSION

In this work, the feasibility of using a swine model to reflect human bioimpedance measurements has been analyzed. The advantage of a swine model would be the possibility to examine the influence of pathologies and certain physiological sources on the ICG signal of humans.

First, it has been shown that the base impedance correlates with the fluid content of the body, and with modified equations the body water content can be calculated. The assessment of TBW or the thoracic fluid content is one way to detect lung edema in heart failure patients. Second, the morphology of the ICG signal is similar to a human ICG which allows the extraction of established characteristic points and allows the calculation of LVET, SV and CO. CO trends are similar but CO is underestimated by calculations using bioimpedance. Hence, algorithms to calculate SV have to be adapted because of different signal amplitudes of ICG and impedance.

All in all, promising results have been obtained. However, more effort must be made in creating new algorithms for swine models and more pigs should be measured to get representative results.

ACKNOWLEDGMENT

This work was partly funded by “HeartCycle”, an EU-project about compliance and effectiveness in Heart Failure (HF) and Coronary Heart Disease (CHD) closed loop management and has been supported by Philips Research Europe.

In addition, other parts were funded by the German Federal Ministry of Science and Education, the State of North Rhine-Westphalia (NRW, Germany, Ziel2 program).

REFERENCES

[1] G. Cotter, “Impedance cardiography revisited,” *Physiological Measurement*, vol. 27, pp. 817–827, 2006.
[2] C. Gabriel, “The dielectric properties of biological tissues,” *Physics in Medicine and Biology*, vol. 41, pp. 2231–2249, 1996.
[3] M. Packer, “Utility of impedance cardiography for the identification of short-term risk of clinical decompensation in stable patients with chronic heart failure,” *Journal of the American College of Cardiology*, vol. 47, pp. 2245–2252, 2006.
[4] J. Van De Water, “Impedance cardiography—the next vital sign technology?,” *Chest*, vol. 123, pp. 2028–2033, 2003.
[5] T. Debski, “Stability of cardiac impedance measures: Aortic opening (b-point) detection and scoring,” *Biological Psychology*, vol. 36, pp. 63–74, 1993.
[6] P. Carvalho, “Robust characteristic points for ICG: Definition and comparative analysis,” *Biosignals—International Conference on Bio-inspired Systems and Signal Processing*, 2011.
[7] R. Kushner, “Estimation of total body water by bioelectrical impedance analysis,” *The American Journal of Clinical Nutrition*, vol. 44, pp. 417–424, 1986.

Lecture Notes on Impedance Spectroscopy, Volume 4 – Kanoun (ed)
© 2014 Taylor & Francis Group, London, ISBN 978-1-138-00140-4

Impedance spectroscopy study of healthy and atelectatic lung segments of mini pigs

S. Aguiar Santos
Philips Chair for Medical Information Technology (MedIT), RWTH Aachen University, Aachen, Germany

M. Czaplik
Philips Chair for Medical Information Technology (MedIT), RWTH Aachen University, Aachen, Germany
Department of Anesthesiology, University Hospital RWTH Aachen, Aachen, Germany

S. Leonhardt
Philips Chair for Medical Information Technology (MedIT), RWTH Aachen University, Aachen, Germany

ABSTRACT: In this study, bioimpedance spectroscopy measurements ranging from 1 kHz to 1 MHz were made on *post mortem* lung segments of mini pigs. Healthy and partly atelectatic lungs injured by mechanical ventilation were used, both deflated and under air inflation. The results, presented on a Cole-diagram of the complex impedance, conductance by frequency and conductance differences at different frequency plots, are discussed. These data confirm that detection of certain lung diseases may be possible based on differences in impedance measured at different frequencies.

Keywords: bioimpedance, dielectric properties, lung diseases

1 INTRODUCTION

In Europe, respiratory diseases rank second (after cardiovascular diseases (CVDs)) with regard to mortality, incidence, prevalence and healthcare costs [1]. Therefore, lung monitoring for the detection and prevention of lung diseases, such as atelectasis, pneumonia, or pulmonary edema, is important. For detection of these diseases various methods are available. The classical techniques include X-ray imaging, sonography, computed tomography (CT), bronchoscopy, blood sample analysis, and estimation of extravascular lung water by thermodilution. However, some of these methods are invasive and/or involve the use of catheters, and the non-invasive methods have various disadvantages, such as poor resolution in the case of X-rays, or ionizing radiation and the impossibility of bedside use in case of CT. In addition, mechanical ventilation of patients with lung disease requires an individual configuration of the ventilatory parameters. Moreover, respiratory parameters need to be continuously evaluated since ventilation itself can lead to ventilator-induced lung injury (VILI). Therefore, it is important to find new methods and devices that help to control mechanical ventilation for the detection and prevention of these diseases.

Since the morphology of a pathologic lung differs from that of a healthy lung, its electrical properties probably also differ. Therefore, the present study analyses the dielectric properties of the lung.

2 DIELECTRIC PROPERTIES OF TISSUE AND BIOIMPEDANCE SPECTROSCOPY

The dielectric properties of human tissue have been extensively studied. Whereas some groups refer to one particular tissue, e.g. skin [2], Gabriel et al. (1996) created a database of a wide variety of biological materials [3–6], including an electronic database [7]. Although this database is useful and used by many research groups, it is not complete and has several limitations, especially at frequencies below 1 MHz, as mentioned by the authors themselves [6].

The dielectric properties of tissues can be determined by, e.g., measuring their impedance. Bioimpedance spectroscopy performs measurements in a range of frequencies. This technique employs four electrodes: two outer electrodes for current injection, and two inner ones for voltage measurement. These measurements are usually performed in the β-dispersion region (1 kHz–1 MHz) [8], where mainly the capacitance of the individual cell envelope and the resistivity of the cell interior are measured [9]. Since few data on β-dispersion of lung tissue are available, and our interest is focused on detection and prevention of lung diseases, e.g., pulmonary edema, this is an important study. With better knowledge of these properties, medical techniques such as Electrical Impedance Tomography (EIT) devices might be improved.

3 MATERIALS AND METHODS

3.1 *Postmortem lungs*

After euthanasia of 5 mini pigs (weight 30–35 kg) lungs were immediately removed. Although no direct harm to the lungs was applied during the prior experiments, lungs were more or less impaired by recent mechanical ventilation (in particular atelectasis in dependent regions). Next, lungs were examined macroscopically, diverse segments were separated and then classified by an experienced physician regarding injury. Figure 1 shows a schematic of the lung segments.

A segment from each pig was used (S1–S5, Fig. 3). Healthy lungs (or with almost no atelectasis), as well as lungs partly collapsed and with hemorrhagic regions (atelectatic), were measured both deflated and under air inflation; for this, a manual syringe (20 mL) was used (Fig. 2). Air volume was adapted to the segment size in order to achieve adequate volume extension.

3.2 *Measurement scenario*

The measurements were performed inside a small plastic box, equipped with 6 electrodes of stainless steel (V2A) which were 0.07 mm thick, 13 mm wide, separated equidistantly

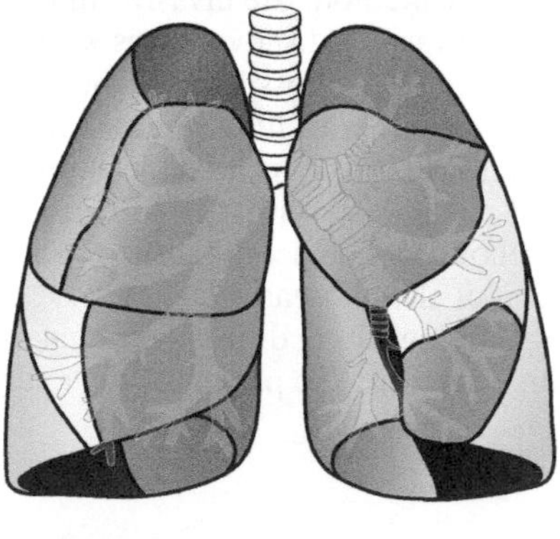

Figure 1. Schematic of the lungs showing their segments.

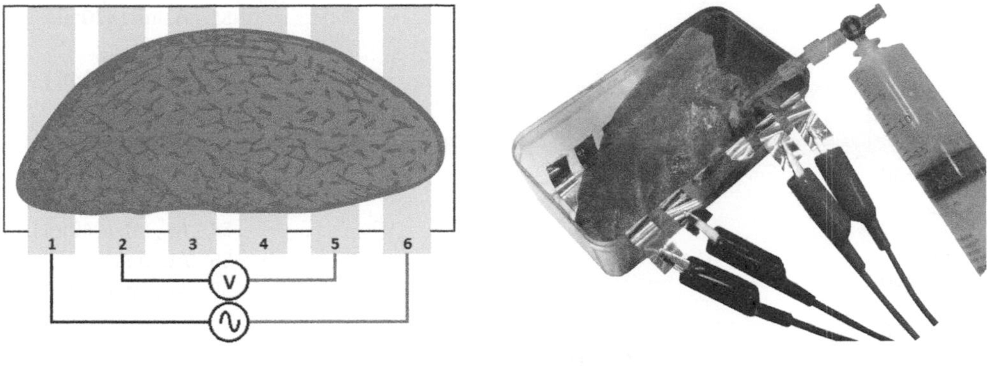

1
2
3
4
5
6
V

Figure 2. Measurement scenario: Left, a schematic of the set-up; right, the real scenario.

by 7 mm (Fig. 2). The outer electrodes (1, 6) were connected to the current output of a Precision LCR Meter (Agilent E4980A, Agilent Technologies), and two middle inner (2, 5) to the input measurement. The two remaining inner electrodes (3, 4) were left unconnected.

For each segment, alternating electrical currents at 200 frequencies in the range of 1 kHz to 1 MHz were injected. The complex impedance Z [resistance R (real) and reactance X (imaginary)] of the lung tissue was measured directly. Posteriorly, the conductance G was estimated ($G[S] \sim 1/Re\{Z\}$).

4 RESULTS

Figure 3 shows the Cole-diagram of the complex impedance. The results of the segments of the five pigs (S1–S5) are plotted. For each case, the measurements were performed twice to avoid artefacts. As no significant differences were found, only one measurement of each case was plotted. Because lung tissue S3 was leaking, only 'deflated' measurements could be performed. At first sight the curves differ significantly, particularly with regard to the amplitudes. On further analysis, we found that the curve characteristic seems to be associated with the lung health status. The segments with atelectasis show a more accentuated curve, whereas the healthy ones present a quasi linear curve.

Figure 4 shows the estimated conductances of the lung segments depending on the frequency used. Again, there is a difference in the curves related to the lung health status. The partly collapsed and hemorrhagic regions show a more accentuated curve, increasing in conductance with higher frequency. This behavior can be explained by comparing the conductivities of the lung and blood as provided by the database in [7], and shown in Figure 5. Regarding blood, conductivity is constant up to 100 kHz (about 2–2.5 times higher than the deflated lung) after which it increases up to 1 MHz. In contrast, the conductivity curve of a deflated lung is constantly increasing.

Therefore, for hemorrhagic impaired lung regions, a constantly increased conductivity is expected up to 100 kHz. At a higher frequency, a more accentuated rise should appear. This is in accordance with our results. An exception was the segment S3; here atelectasis was also macroscopically less than in S4 and S5. However, the evaluation of lung morphology was based on macroscopic observation only. Moreover, we observed a reduced conductance in inflated lung segments, which is also in accordance with the data presented by [7]. A reason for this is the low conductivity of air (3×10^{-15} to 8×10^{-15} S/m, at 20°C).

Figure 6 shows the conductance difference in lung segments between frequencies 100 kHz–1 MHz. Also here there is a marked difference between healthy and atelectatic lung segments, except for the different behavior of segment S3 (as explained above).

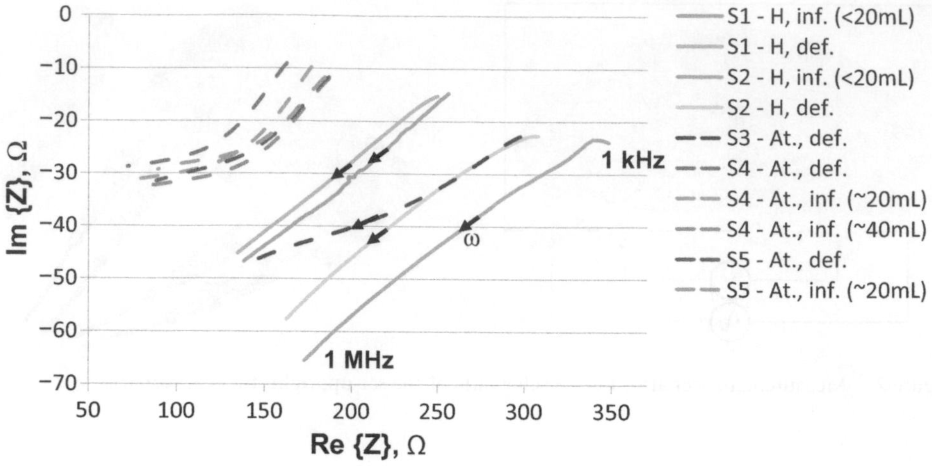
0
−10
−20
−30
−40
−50
−60
−70
Im {Z}, Ω
50
100
150
200
250
300
350
Re {Z}, Ω
1 kHz
1 MHz
ω
S1 - H, inf. (<20mL)
S1 - H, def.
S2 - H, inf. (<20mL)
S2 - H, def.
S3 - At., def.
S4 - At., def.
S4 - At., inf. (~20mL)
S4 - At., inf. (~40mL)
S5 - At., def.
S5 - At., inf. (~20mL)

Figure 3. Cole-diagram of the measured complex impedance (H, healthy or almost no atelectasis; At, Atelectasis; inf, inflated; def, deflated).

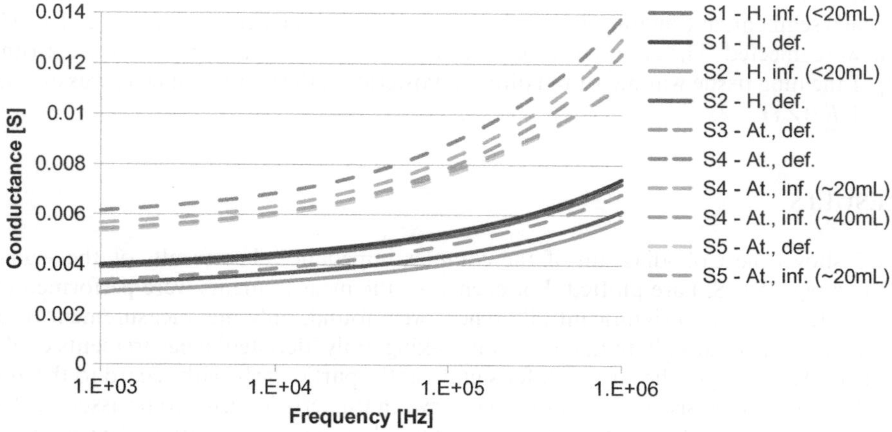
0.014
0.012
0.01
0.008
0.006
0.004
0.002
0
Conductance [S]
1.E+03
1.E+04
1.E+05
1.E+06
Frequency [Hz]
S1 - H, inf. (<20mL)
S1 - H, def.
S2 - H, inf. (<20mL)
S2 - H, def.
S3 - At., def.
S4 - At., def.
S4 - At., inf. (~20mL)
S4 - At., inf. (~40mL)
S5 - At., def.
S5 - At., inf. (~20mL)

Figure 4. Estimated conductances (see section 3.2) of lung segments depending on the frequency used (1 kHz–1 MHz). (H, healthy or almost no atelectasis; At, Atelectasis; inf, inflated; def, deflated).

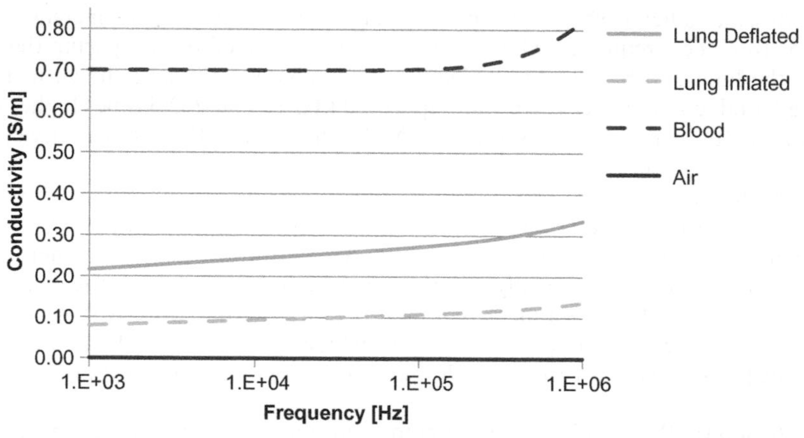
0.80
0.70
0.60
0.50
0.40
0.30
0.20
0.10
0.00
Conductivity [S/m]
1.E+03
1.E+04
1.E+05
1.E+06
Frequency [Hz]
Lung Deflated
Lung Inflated
Blood
Air

Figure 5. Conductivity data from an inflated and deflated lung, and from blood and air: reproduced from [7].

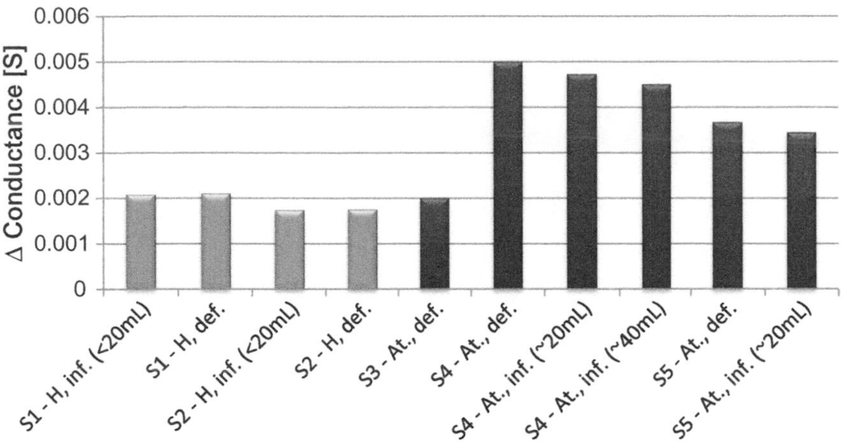

0.006
0.005
0.004
0.003
0.002
0.001
0
Δ Conductance [S]
S1 - H, inf. (<20mL)
S1 - H, def.
S2 - H, inf. (<20mL)
S2 - H, def.
S3 - At., def.
S4 - At., def.
S4 - At., inf. (~20mL)
S4 - At., inf. (~40mL)
S5 - At., def.
S5 - At., inf. (~20mL)

Figure 6. Conductance difference between the 100 kHz and 1 MHz frequencies (H, healthy or almost no atelectasis; At, Atelectasis; inf, inflated; def, deflated).

5 CONCLUSION

In summary, dielectric properties differ between healthy and injured lungs due to atelectasis caused by mechanical ventilation. It is shown that detection of more or less collapsed lung tissue was possible by analysing differences in electrical impedanceat different frequencies. This can be a promising feature for multi-frequency EIT devices. More studies are needed to evaluate additional diseases, such as pulmonary edema or pneumonic infiltrates.

ACKNOWLEDGMENT

Susana Aguiar Santos is supported by the PhD Grant SFRH/BD/76441/2011 awarded by the Portuguese Foundation for Science and Technology (FCT), Portugal.

REFERENCES

[1] P. Mladovsky, S. Allin, C. Masseria, C. Hernández-Quevedo, D. McDaid, and E. Mossialos, *Health in the European Union. Trends and analysis*. Copenhagen: World Health Organization, on behalf of the European Observatory on Health Systems and Policies, 2009.

[2] U. Birgersson, E. Birgersson, P. Åberg, I. Nicander, and S. Ollmar, "Non-invasive bioimpedance of intact skin: mathematical modeling and experiments," *Physiological Measurement*, vol. 32, pp. 1–18, Jan. 2011.

[3] S. Gabriel, R.W. Lau, and C. Gabriel, "The dielectric properties of biological tissues: II. measurements in the frequency range 10 hz to 20 GHz," *Physics in Medicine and Biology*, vol. 41, pp. 2251–2269, Nov. 1996.

[4] C. Gabriel, S. Gabriel, and E. Corthout, "The dielectric properties of biological tissues: I. literature survey," *Physics in Medicine and Biology*, vol. 41, pp. 2231–2249, Nov. 1996.

[5] S. Gabriel, R.W. Lau, and C. Gabriel, "The dielectric properties of biological tissues: III. parametric models for the dielectric spectrum of tissues," *Physics in Medicine and Biology*, vol. 41, pp. 2271–2293, Nov. 1996.

[6] C. Gabriel, A. Peyman, and E.H. Grant, "Electrical conductivity of tissue at frequencies below 1 MHz," *Physics in Medicine and Biology*, vol. 54, pp. 4863–4878, Aug. 2009.

[7] C. Gabriel and S. Gabriel, "Compilation of the dielectric properties of body tissues at RF and microwave frequencies," *http://niremf.ifac.cnr.it/docs/DIELECTRIC/home.html*, 1997.

[8] S. Grimnes and O.G. Martinsen, *Bioimpedance and Bioelectricity Basics*. London: Academic, 2008.

[9] H.P. Schwan and C.F. Kay, "The conductivity of living tissues," *Annals of the New York Academy of Sciences*, vol. 65, pp. 1007–1013, Aug. 1957.

Lecture Notes on Impedance Spectroscopy, Volume 4 – Kanoun (ed)
© 2014 Taylor & Francis Group, London, ISBN 978-1-138-00140-4

System for bioimpedance signal simulation from pulsating blood flow in tissues

R. Gordon & K. Pesti
Thomas Johann Seebeck Department of Electronics, Tallinn University of Technology, Estonia

ABSTRACT: Blood pulsation in smaller arteries is modeled in a small tissue sample. The hierarchical vascular network is generated using Constrained Constructive Optimization algorithm and only arterial side of the network is modeled. Pulsation of the arteries is introduced to the model using pressure wave and propagation speed considerations. 2-electrode bioimpedance measurement scenario is simulated. The dynamic change of electric bioimpedance due to blood pulsation in the modeled tissue sample is simulated using multi-frequency Finite Difference Method calculation.

Keywords: bioimpedance, vascular system, simulation, pulse wave velocity, arterial tree, Constrained Constructive Optimization

1 INTRODUCTION

Many electric diagnostic methods are today used in medicine, that can noninvasively estimate the health of organs deep inside the body. The best know methods are electrocardiogram ECG, electroencephalogram EEG and electromyogram EMG. Those are active methods, where the electric signal originates inside the body endogenously. Bioimpedance analysis is a method to evaluate the health of a patient with exogenous electric signals and the results depend on the passive electric properties of the body or organ under study [1]. The passive electric properties of the organ (electric conductivity and dielectric permittivity) depend a lot on blood content that changes dynamically. The electric bioimpedance measurement consists of applying an exogenous electric signal—for example a set voltage at a specific frequency—and measuring the passing electric current due to the electric potential. This can also be done the other way around, introducing a fixed current and measuring the voltage drop. The amplitude and phase information of the measured current (or voltage) is then used to calculate the electric bioimpedance [2].

The main reason that we do research on bioimpedance in this project is to determine the passive electric properties of tissues and see if the dynamics can influence the electric diagnostic methods that are already used. Bioimpedance measurement has been used as a method for diagnostics in its own right. For example monitoring the breathing activity with chest bioimpedance measurement (conductivity of the lungs changes significantly with air content in the lungs [3, 4]), cardiac dynamics has been monitored to determine the stroke volume noninvasively [5, 6] and transplanted muscle flaps have been monitored using bioimpedance to determine possible tissue edema and post-operation complications [7, 8].

The simulation of electric impedance signal in a tissue sample is made here to investigate the role of tissue dynamics in various electric diagnostic methods in medicine. Most electric diagnostic methods apply electrodes to the patient to measure the electric signals. If the whole patient would be a source of the signal homogenously, then the majority of the measured signal would originate in close proximity of the electrodes. The measurement is most sensitive to those areas. The term half sensitivity volume [9] has been used to describe the region, where 50% of the signal would come from. This region is typically a small area under the measurement electrodes. The location and extent of that half sensitivity volume is somewhat similar with bioelectric signals of endogenous origin as well as impedance signals with external excitation. Both types of measurements depend on the signal source as well as the conductivity of the whole media. Therefore we aim to quantify how much the local dynamics of the vascular system could influence the measurement of those signals.

The vascular system has been the subject of modeling for a long time. Most proliferating are one-dimensional models of arteries, which also exhibit dynamics of blood pulsation [2]. Those models are all limited to larger arteries and only approach arteries as small as 1 mm diameter. Arteries narrower than that have not been addressed in fully dynamic vascular system models. Smaller arteries, arterioles and capillaries are rarely modeled with dynamics—especially dynamics that include blood volume change that would translates to changes in tissue electric properties. In this approach we model the blood pulsation in smaller arteries in a tissue with dimensions around 10 mm and artery diameters that start from 17.4 μm and go down to capillary level. The multi-frequency electric bioimpedance signal due to blood pulsation in the tissue is simulated after the dynamics in the modeling of the dynamic micro-scale arterial network. The authors have not found any previous work with a dynamic computer-model of the system of small arteries that includes bioimpedance signal measurement simulation.

2 METHODS

The model of arterial network anatomy was obtained with automatic vascular network generating algorithm based on Constrained Constructive Optimization (CCO) principle [10]. The CCO principle works by inserting a random point into a prescribed volume and by drawing a new arterial segment to the randomly placed point from the already existing arterial tree. It finds the best contact segment for the new point and then finds the best location for the new cross-section to minimize the vascular system total volume while keeping a desired blood-flow. The algorithm working on this principle is able to generate vascular networks very similar to the anatomical ones [10]. In this work the CCO principle was used in MATLAB to generate an arterial network of 3999 total segments. The network had 2000 arterial end-segments, which feed the tissue with random placement of end-points.

The feeding artery had around 17 μm diameter and all consecutive segments had diameters somewhat smaller, optimized for best flow with minimal arterial tree total volume as described above (and in more detail in [10]). The algorithm works for 3D arterial system generation as well as 2D generation for specialized cases, like retina and other membranes. The 2D version was used in this initial study for simplicity, fast results and illustrative purposes (Figure 1, Figure 2). The 2D version of the vascular tree was generated with the same CCO algorithm but in a tissue volume area with dimensions it 10 mm × 10 mm × 0.28 mm. This choice forced the algorithm to create a 2D tree leaving no room for vascular segments to above/below each other. The 2D choice was made to allow for single layer Finite Difference Method calculation.

After the generation of static arterial tree model dynamics was introduced to the network that mimics pulse wave propagation in the tissue with each heart-beat. Then impedance electrodes were applied to both sides of the model and Finite Difference Method simulation of impedance was made for each time-slice of the dynamic model (total 60 time-slices for 1 second).

The dynamics of pulsatile blood flow were implemented with a 10% radius increase with the duration of half a second. The shape of the pulse was a simple phase-shifted cosine function (Figure 3). The shape and extent of the dynamics was chosen for simplicity and testing purposes. In reality the distensibility of the vessels is getting smaller with decreasing vessel diameter and increasing stiffness [10]. The heart rate was considered to be 60 bpm and simulation time was one heartbeat with 60 simulation time-steps during the one second period. The propagation of the pulse was modeled with 5 m/s pulse wave velocity after the onset (entrance) of the pulse in the feeding artery (left-center of the model on Figure 2). Real pulse

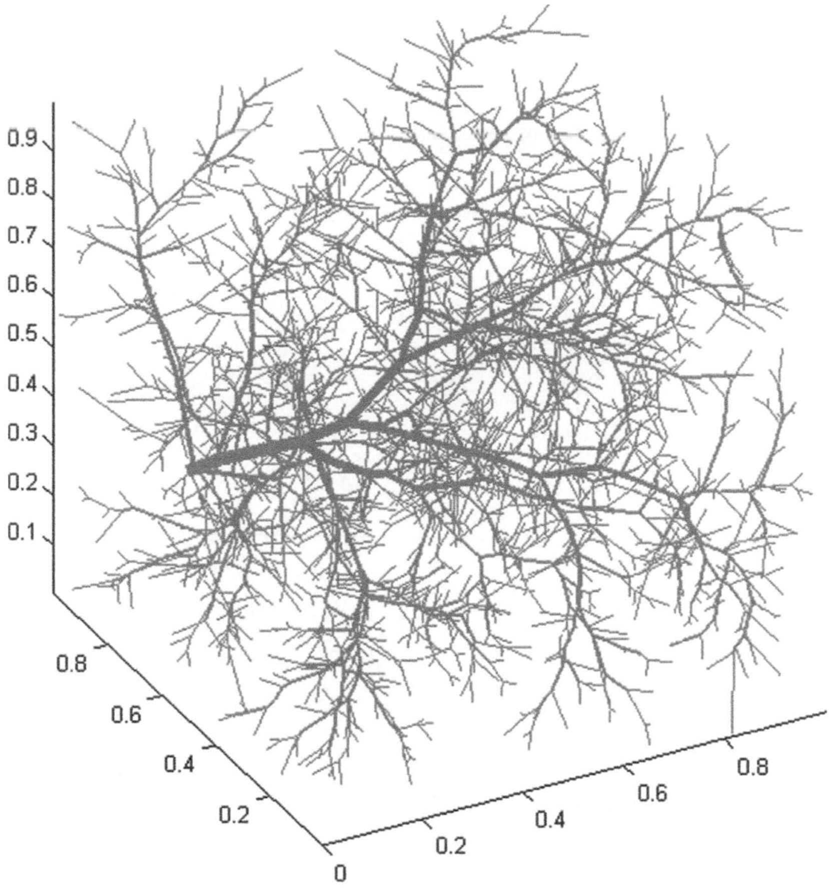
0.9
0.8
0.7
0.6
0.5
0.4
0.3
0.2
0.1
0.8
0.6
0.4
0.2
0
0.2
0.4
0.6
0.8

Figure 1. Automatically generated 3D arterial tree model.

wave velocity in small arterioles and capillaries should be much faster in real muscle or adipose tissues and the speed should be increasing due to the increase of vascular stiffness with thinning diameter. But for the initial testing of the system simplifications were made in this area. Pulse wave velocity in blood vessels is orders of magnitude higher than average blood flow speed, which is in the range of 3 to 5 mm/s in capillaries [11]. Each segment (short section of straight artery between two consecutive bifurcations) of the total 3999 segments was given the 10% amplitude increase with cosine function shape full period but phase shifted and timed with a short delay proportional to the length of arterial tree between that segment and the root segment (and considering the 5 m/s propagation speed). This attributed to a pulse wave entering the root segment and propagating through the tree towards all 2000 end-points. The widening of some of the segments during a heart-beat is illustrated in Figure 3. The segment number in Figure 3 does not directly indicate the distance of the segment from the root. It represents the order of the segment birth during the CCO tree generation. Some segments that are generated later in the process may end up being close to the root segment. Segment no 1 is the first one but all others after that have different distances from the root and various diameters and flows, not in numerical order.

The conductivities of the tissues (muscle with pulsating blood content, Table 1) were obtained from the internet database of dielectric properties of body tissues [12–14]. Three frequencies—1 kHz, 10 kHz and 100 kHz—were chosen for this initial simulation run

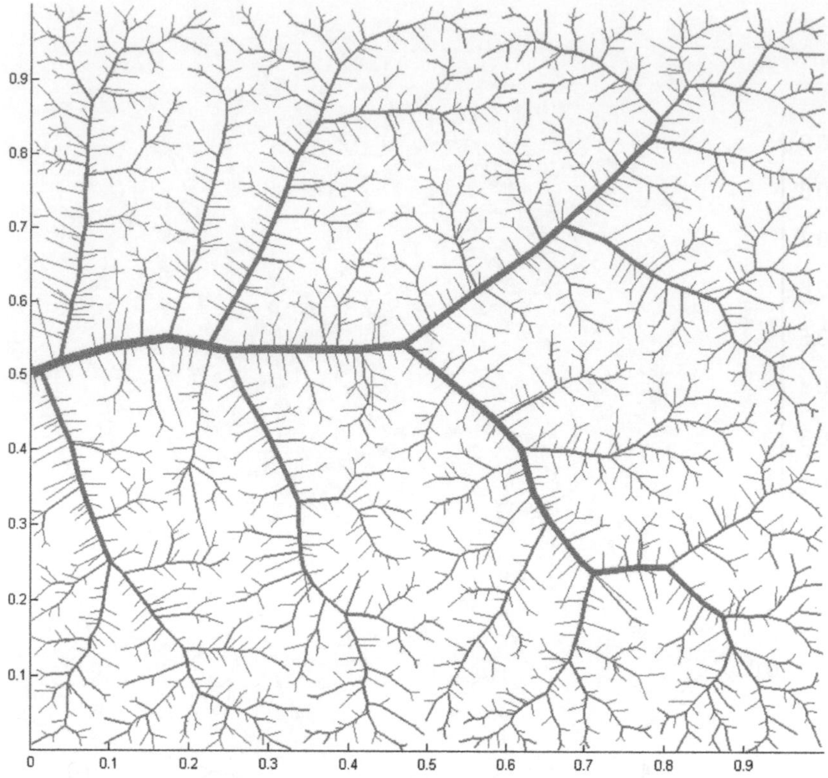
0.9
0.8
0.7
0.6
0.5
0.4
0.3
0.2
0.1
0
0.1
0.2
0.3
0.4
0.5
0.6
0.7
0.8
0.9

Figure 2. 2D version of the generated tree, which was used in dynamics simulation.

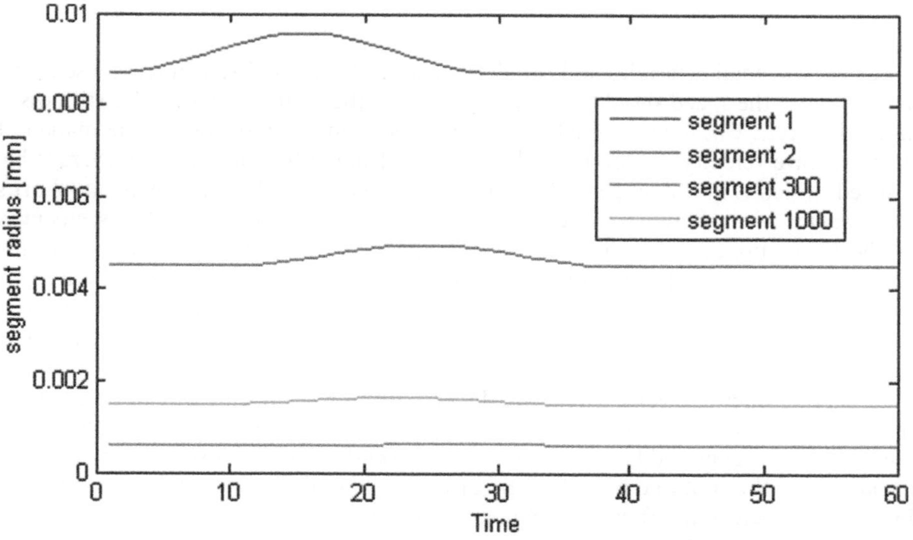
segment radius [mm]
0.01
0.008
0.006
0.004
0.002
0
0
10
20
30
40
50
60
Time
segment 1
segment 2
segment 300
segment 1000

Figure 3. Dynamics of pulsating flow implemented to the segments of arterial tree.

Table 1. Conductivities of tissues used in simulation.

Frequency	Conductivity of muscle S/m	Conductivity of blood S/m
1 kHz	0.32115	0.7
10 kHz	0.34083	0.70004
100 kHz	0.36185	0.70292

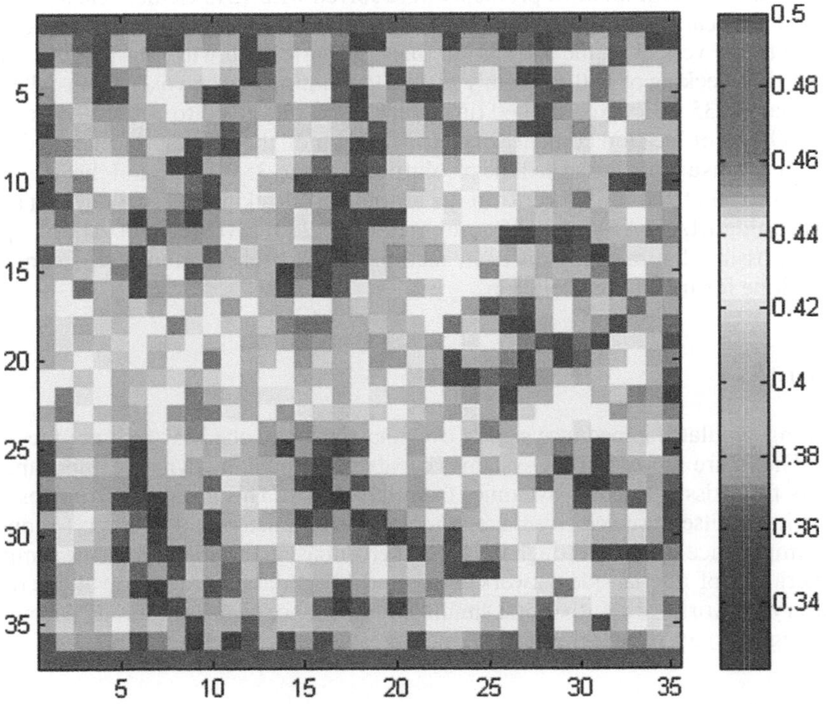

5
10
15
20
25
30
35
5
10
15
20
25
30
35
0.5
0.48
0.46
0.44
0.42
0.4
0.38
0.36
0.34

Figure 4. Conductivity map for impedance simulation. Electrodes are positioned on top and on the bottom of the 35 × 35 voxel tissue map.

although the full frequency characteristic could be used if necessary. Dielectric permittivity data was neglected and only conductivity was used for simplicity in this test. Although the Finite Difference Method implementation in MATLAB® is fully capable of calculations with complex numbers and the dielectric properties can be used if necessary.

The impedance simulation of the dynamics was carried out with Finite Difference Method (FDM) implemented in MATLAB. Alternatively, the discretization of equations can be performed by use of Finite Element Method (FEM), wavelets techniques, etc. [15,16]. Electrodes were positioned on 2 sides of the 10 mm square tissue model, current was inserted and voltages were calculated for each case of 60 time-steps. Electric impedance maps with resolution 35 × 35 × 1 were derived from the dynamic vascular network (Figure 4).

The blood content (dynamic over 60 time-steps) in each voxel was obtained from the 3999 segment vascular network with Monte-Carlo method. The task was to find the overlapping volume of each voxel (cube) with each segment (cylinder). Function for finding the intersecting volume of a cube with an arbitrarily positioned cylinder was not readily available.

Therefore we resolved to Monte-Carlo method to find the intersecting volume. 500 randomly placed points for each voxel were used to check the existence of blood vessel. This gave a statistically good indication of the volume of blood vessels for this voxel. The method allowed to produce volume data with the resolution of $35 \times 35 \times 1$ about percentage value of how much blood is in each voxel instead of bulk tissue.

Conductivities of muscle and blood (table 1) were then used with the tissue volume data to produce a conductivity image with $35 \times 35 \times 1$ resolution for FDM calculation. The conductivity image was generated for each of the 60 time-steps of dynamics and 3 frequencies. FDM calculation to find electrode-to-electrode impedance measurement value was performed for each time-step and frequency.

All together 180 FDM forward problems were solved with 1295 element each. The resolution of the FDM calculation could have been significantly higher but the most time consuming was the blood vessel volume calculation that was performed with Monte-Carlo method. It required the checking of 500 randomly placed locations for each voxel against 3999 vessel segments, having 35×35 voxels and 60 time-steps. This amounted to $500 \times 3999 \times 35 \times 35 =$ 2'449'387'500 point location comparisons. This section of the whole project ended up to be the most time-consuming. We scaled the resolution down to $35 \times 35 \times 1$ and the Monte-Carlo points down to 500 to be able to perform the automated checking in just 3 days on a fast PC in 2013 (running MATLAB on single core/thread at 3.2 GHz frequency). Minor optimizations were possible but radically different methods and parallelization of the computation should be done for increased resolution.

3 RESULTS

The resulting simulated impedance signals over the course of one heartbeat are presented on Figure 5. The figure shows that the system is capable of simulating dynamic impedance signal that results from tissue vascular dynamics from heart-beat. This gives opportunities to carry out research for pulse-wave-velocity effects, rheological parameters and tissue perfusion estimation from impedance signal. Although the blood percentage in this modeled tissue sample (only limited hierarchy of arterial side network) is around 1% and the introduced volume dynamics of the artery network is 10%, the total simulated impedance signal amplitude becomes more than 1% (Figure 5) and would therefore be measureable by dedicated equipment.

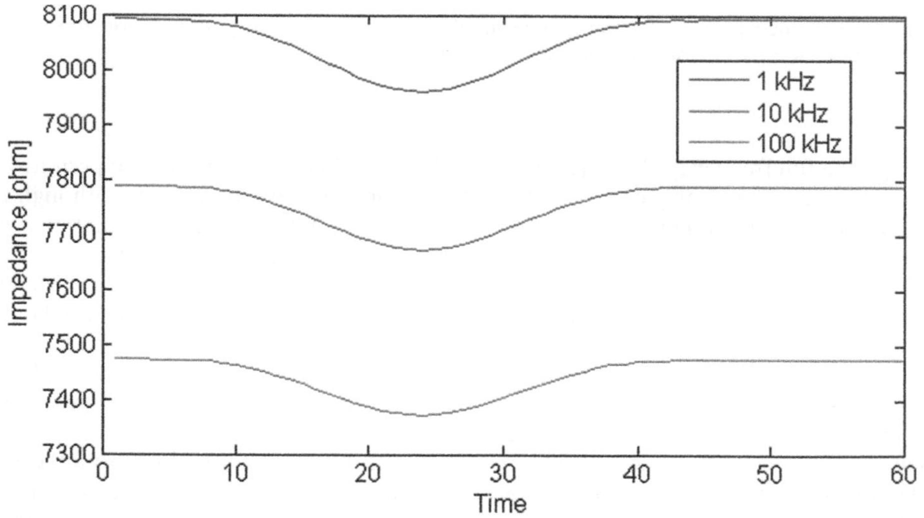
8100
8000
7900
7800
7700
7600
7500
7400
7300
Impedance [ohm]
0
10
20
30
40
50
60
Time
1 kHz
10 kHz
100 kHz

Figure 5. Impedance signals simulated with the 10 mm square dynamic tissue sample between 10 mm long electrodes on both sides.

4 DISCUSSION

This model is created for research in medical electric diagnostic methods, where the tissues locally are expected to have significant contribution to the measured signal. They include bioimpedance methods with electrodes, because the increased sensitivity under the electrodes, as well as eddy-current methods, where signal sensitivity depends heavily on the distance from the measurement coils. As the diagnosed organs and areas of interest are often away from the electrodes or coils, the tissues closer to the sensors can have unexpectedly large contribution to the measured signals.

When the tissue modeling is performed with actual parameters of human live tissues, the effect of the vascular dynamics in the local tissue can be revealed. Local effects like temperature related vasodilation and vasoconstriction in skin and body temperature balance could also be assessed with the help of vascular dynamics modeling.

The resolution of the conductivity image (Figure 4) is still a good representation for FDM impedance calculation because each voxel is a statistically significant presenter of the muscle/blood percentage and therefore the changing conductivity. Still much improved methods for calculation have to be found for 3D capability and for more complex organ or body simulation scenarios. The good enough resolution of the voxel data and Monte-Carlo calculation is also indicated by the smoothness of the simulated signals that is apparently free of any statistical noise.

This simulation study is the first publication of this tissue dynamics impedance simulation system and demonstrates the functioning of the simulation system. In the future the system will be used for simulating real measurement scenarios to reveal if local and small scale vascular dynamics in tissues really influence medical electric diagnostic methods. The authors hope that the knowledge of small scale impedance dynamics in the tissues can reveal innovative diagnostic opportunities.

ACKNOWLEDGEMENT

The authors thank European Union Structural Funds in Estonia (project 1.0101.01-0480), and the Estonian Science Foundation (grant 9394) for the support.

REFERENCES

[1] O. Martinsen and S. Grimnes, *Bioimpedance and bioelectricity basics*. Academic press, 2011.

[2] P. Reymond, F. Merenda, F. Perren, D. Rüfenacht, and N. Stergiopulos, "Validation of a one-dimensional model of the systemic arterial tree," *American Journal of Physiology-Heart and Circulatory Physiology*, vol. 297, no. 1, pp. H208–H222, 2009.

[3] I. Frerichs, "Electrical impedance tomography (eit) in applications related to lung and ventilation: A review of experimental and clinical activities," *Physiological Measurement*, vol. 21, no. 2, p. R1, 2000.

[4] A. Adler, R. Amyot, R. Guardo, J. Bates, and Y. Berthiaume, "Monitoring changes in lung air and liquid volumes with electrical impedance tomography," *Journal of Applied Physiology*, vol. 83, no. 5, pp. 1762–1767, 1997.

[5] W. Kubicek, R. Patterson, and D. Witsoe, "Impedance cardiography as a noninvasive method of monitoring cardiac function and other parameters of the cardiovascular system," *Annals of the New York Academy of Sciences*, vol. 170, no. 2, pp. 724–732, 1970.

[6] W. Kubicek, "Impedance plethysmograph," 1967.

[7] S. Kun and R. Peura, "Tissue ischemia detection using impedance spectroscopy," in *Engineering in Medicine and Biology Society, 1994. Engineering Advances: New Opportunities for Biomedical Engineers. Proceedings of the 16th Annual International Conference of the IEEE*, pp. 868–869, IEEE, 1994.

[8] P. Annus, A. Kuusik, R. Land, E. Haldre, M. Min, T. Parve, and G. Poola, "An energy efficient wearable tissue monitor," in *13th International Conference on Electrical Bioimpedance and the 8th Conference on Electrical Impedance Tomography*, pp. 240–243, Springer, 2007.

[9] K. Wendel, N.G. Narra, M. Hannula, J. Hyttinen, and J. Malmivuo, "The influence of electrode size on eeg lead field sensitivity distributions," *Intl. J. of Bioelectromag*, vol. 9, no. 2, pp. 116–117, 2007.

[10] R. Karch, F. Neumann, M. Neumann, and W. Schreiner, "A three-dimensional model for arterial tree representation, generated by constrained constructive optimization," *Computers in biology and medicine*, vol. 29, no. 1, pp. 19–38, 1999.

[11] J. Li, *Dynamics of the vascular system*, vol. 1. World Scientific Publishing Company, 2004.

[12] C. Gabriel, S. Gabriel, and E. Corthout, "The dielectric properties of biological tissues: I. literature survey," *Physics in medicine and biology*, vol. 41, no. 11, pp. 2231–2249, 1999.

[13] S. Gabriel, R. Lau, and C. Gabriel, "The dielectric properties of biological tissues: II. measurements in the frequency range 10 hz to 20 ghz," *Physics in medicine and biology*, vol. 41, no. 11, pp. 2251–2269, 1999.

[14] S. Gabriel, R. Lau, and C. Gabriel, "The dielectric properties of biological tissues: III. Parametric models for the dielectric spectrum of tissues," *Physics in medicine and biology*, vol. 41, no. 11, pp. 2271–2293, 1999.

[15] J. Majak, M. Pohlak, M. Eerme, and T. Lepikult, "Weak formulation based haar wavelet method for solving differential equations," *Applied Mathematics and Computation*, vol. 211, no. 2, pp. 488–494, 2009.

[16] J. Majak, M. Pohlak, and M. Eerme, "Application of the haar wavelet based discretization technique to orthotropic plate and shell problems," *Mechanics of Composite Materials*, vol. 45, pp. 631–642, 2009.

Lecture Notes on Impedance Spectroscopy, Volume 4 – Kanoun (ed)
© 2014 Taylor & Francis Group, London, ISBN 978-1-138-00140-4

Impedance-based infection model of human neutrophils

A. Schröter
Institute of Solid State Electronics, Technische Universität Dresden, Dresden, Germany

A. Rösen-Wolff
Department of Pediatrics, University Hospital Carl Gustav Carus, Dresden, Germany

G. Gerlach
Institute of Solid State Electronics, Technische Universität Dresden, Dresden, Germany

ABSTRACT: Neutrophil granulocytes play an important role in the human immune defence. One of their weapons against bacteria and fungi is the release of Neutrophil Extracellular Traps (NETs) which form in great amounts a biofilm. Electrical Impedance Spectroscopy (EIS) enables the examination of the NET-formation to determine the reaction parameters and to investigate the state of wound inflammation. In this work we analyze the behaviour of neutrophils *in vitro* and develop an empiric model of the states of NET-formation. This model can be used to develop a wound sensor which monitors the NET-formation caused by an infection *in vivo*.

Keywords: Neutrophil Extracellular Traps, bioimpedance, dielectric cell properties, wound healing

1 INTRODUCTION

Wound care has an outstanding relevance in clinical care. Reasons are the high incidence of chronic wounds and linked with it high therapeutic investments [1]. Recommended treatments come along with modern hydroactive wound covers which contain hydrogels, hydrocolloids and foams [2]. They extend the retention time of wound covers which also enhances wound healing. This advantage is rarely used, especially because of the possible complication of an infection. However, this problem could be solved by measuring the infection status of a wound with a sensor.

An infection goes by with a lot of complex biological processes, e.g. the invasion of leukocytes, especially the neutrophil granulocytes. From sensory perspective granulocytes offer an interesting infection mechanism, the creation of Neutrophil Extracellular Traps (NETs). These NETs originate from the cell core which is dissolved when neutrophils sense a harmful microbe. After linking with bactericide granula in the intracellular space the dsDNA is released by the dying neutrophil. This immune reaction is inherent and activated immediately after pathogen contact [3]. Figure 1 shows a fluorescence image of stained neutrophils in several stages of NET-formation. Lobed cell cores of neutrophils can be easily distinguished from haze-like NET fibres.

2 IMPEDANCE MODELS OF CELLS AND DNA

Cells and DNA have a special behaviour when exposing them to alternating electric fields. One possibility to characterize their passive electrical properties is to apply impedance spectroscopic investigations. Schwan [4] was one of the pioneers in impedance measurements on

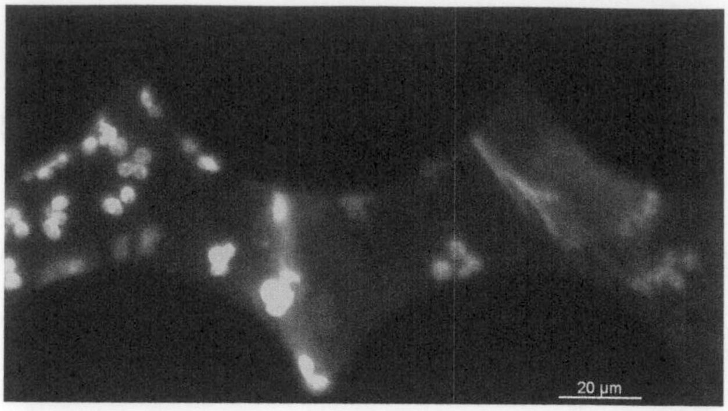
20 µm

Figure 1. Fluorescence image of cultivated neutrophils on electrodes which in the left have a native shaped cell core and in the right release NETs (magnification 40x, DNA-specific SytoxGreen dye).

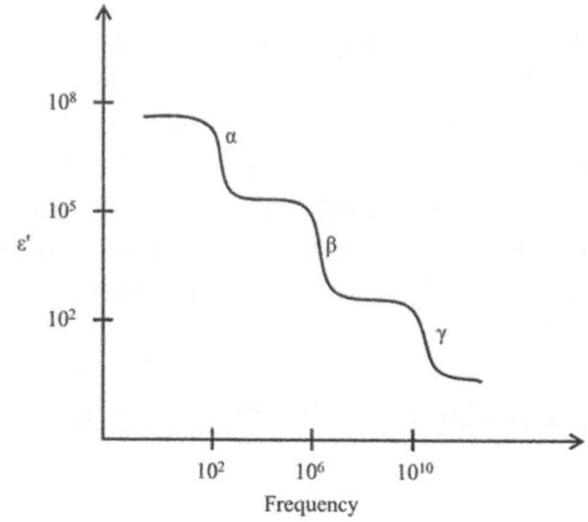
10⁸
10⁵
10²
ε'
α
β
γ
10²
10⁶
10¹⁰
Frequency

Figure 2. Exemplary permittivity spectrum (real part of ε') of cells with the permittivity steps of the α-, β- and γ-dispersion (reprinted from [5]).

cells. He used the complex relative permittivity as a measure for their impedance behaviour. In the spectrum of the relative permittivity of cells permittivity steps—called dispersions—can be found. Namely, the α-, β- and γ-dispersion are relevant for cell measurements. The graph in Figure 2 shows how these steps could be distributed in the frequency spectrum. Every permittivity step has a relation to a physical structure effect. The α-dispersion is caused by charge accumulation effects and active ion transport at the cell membrane. Capacitive effects on barrier layers like cell and organelle membranes are described with one or more β-dispersions. In the higher frequency range dipole effects of water and proteins cause the γ-dispersion.

For a sensor application only frequencies up to the kHz-range are relevant because in higher frequency ranges interfering signals from other sources are too dominant. When measuring below 1 MHz α- and β-dispersion are the predominant effects. The permittivity step of the β-dispersion can be calculated from the cell radius R, the volume fraction p of the sedimented cells in cell culture, the capacitance C_m of the cell membrane and the permittivity ε_0 of free space [6]:

$$\Delta\varepsilon' = \frac{9pRC_m}{4\varepsilon_0} \tag{1}$$

Furthermore, the cut-off frequency where the β-dispersion occurs is dependent on the intracellular conductivity σ_i and the extracellular conductivity σ_a:

$$f_c = \frac{1}{2\pi R C_m \left[\left(\frac{1}{\sigma_i}\right)\left(\frac{1}{\sigma_a}\right)\right]} \tag{2}$$

Impedance spectra of NETs can be considered as similar to those of DNA as they consist mainly of dsDNA. Additionally, the attached granula proteins and the irregular distribution of the NETs should be taken into account. Dielectric properties of dsDNA were first studied in [6]. The authors observed a dispersion mainly caused by rotation of several DNA molecules. The cut-off frequency depends on the relative molar mass M of DNA

$$f_c \sim \frac{1}{M^2} \tag{3}$$

At a relative molar mass of 1.3×10^7 a permittivity step of 2400 at a cut-off frequency of 3.1 Hz was found.

3 SENSOR CONCEPT AND PROCESS CHARACTERIZATION

Measuring just the impedance in the wound would lead to several interferences caused by e.g. the temperature and the pH value. However, a change of these parameters also gives a hint for an ongoing infection in the wound. Combining this information would lead to a sufficient specificity of a wound sensor. Miniaturized sensor concepts already exist for all mentioned parameters. Characterization of wounds in their environment is performed for example for pH value and temperature whereas because of the limited availability of reliable wound models the development for impedance is still in progress. Our wound model environment includes cellular constituents of the wound fluid which can be measured with commercial 2-electrode setups (Roche) for cultivated cells. By using a self-made adapter we connect these electrode wells with an impedance analyzer (ScioSpec ISX-3).

Cell preparation includes isolating neutrophil granulocytes from human blood. These cells are dispensed together with the cell medium containing a cell number of 5×10^4 to 2×10^6. The well bottoms as well as the electrodes have to be coated with poly-L-lysine before. The NET-formation of neutrophils was stimulated by the chemical stimulant phorbol 12-myristate 13-actetate (PMA). The measurement took place at frequencies between 100 Hz and 1 MHz.

4 IMPEDANCE MODEL OF NET-RELEASE

Stimulation of neutrophils to NET-formation leads to a specific change in the dielectric properties of the cells. Especially in the 20 kHz range impedance rises up to 50% of the starting level. Figure 3 shows the difference between unstimulated cells and stimulated cells after 3 hours. The increase lasts for 3 to 6 hours. Afterwards, impedance decreases again which we interpret as a decomposition effect of the NETs (e.g. by elastase).

The permittivity curve reveals one dispersion. Table 1 summarizes the measured permittivity step compared with theoretical values of the β-dispersion calculated by using the formula (1) and (2). The cell diameter of neutrophils is between 12 to 15 µm. An average radius $R = 6.75$ µm can be assumed for a rough estimation. The maximum volume fraction p of the neutrophils by expecting a face-centred cubic order reaches a value of 0.74. Actually, values from (0.6…0.7) for pellets and cell suspension were reported [7–9]. A typical value known from literature for the membrane capacitance C_m is 0.01 F/m^2 and for the intracellular conductivity $\sigma_i = 0.05$ S/m [10]. The extracellular conductivity σ_a can be estimated with a higher conductivity of physiological sodium chloride solution ([0.02]S/m). The comparison

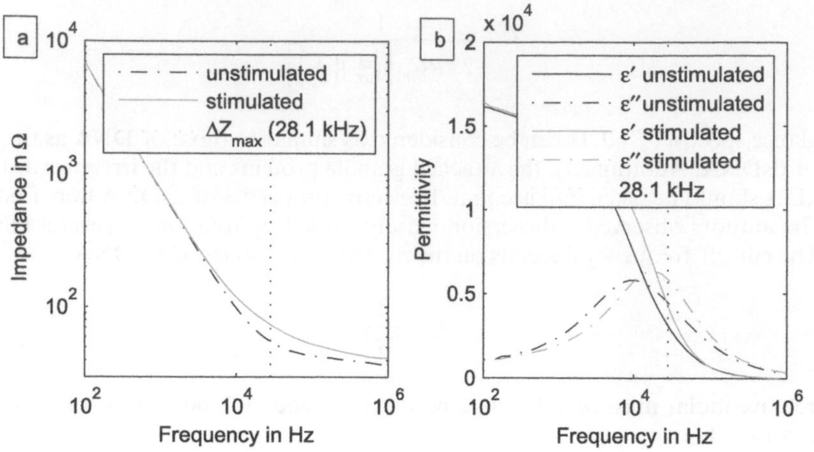
a
b
unstimulated
stimulated
ΔZmax (28.1 kHz)
Impedance in Ω
Frequency in Hz
x 10^4
ε' unstimulated
ε'' unstimulated
ε' stimulated
ε'' stimulated
28.1 kHz
Permittivity
Frequency in Hz

Figure 3. Changes of a) impedance and b) complex permittivity during NET-formation (after 3 hours) compared with unstimulated neutrophils.

Table 1. Calculated and measured values for the permittivity step and cut-off frequency of the β-dispersion.

	Calculated	Measured
$\Delta\varepsilon'$	10.297…12.013	16.000
f_c in Hz	>4.761	6.321

of calculated and measured values shows a slight deviation in the height of the permittivity step and the cut-off frequency. These deviations are a hint for another effect during impedance evaluation of NET-formation, but the destruction of the cell membrane. We assume that the massive appearance of dsDNA are the reason for it.

5 CONCLUSION

We found that impedance measurement is an appropriate method to monitor NET-formation. The investigated dielectric behaviour of neutrophils revealed one dispersion in the kHz range. High changes of the impedance are not just reasoned by the dissolving cell membrane in the β-dispersion but also by other effects like NET-formation. The effect of the DNA-biofilm formed by NETs cannot be estimated up to now because of the lack of knowledge about the amount of released dsDNA. To quantify the NETs a parallel measurement is necessary. This can be achieved by the standard method of NET quantification: The microscopic investigation of fluorescently marked NETs. Further measurements are planned to address this issue. Improved impedance measurements with a finer resolution in the range of the identified dispersion might reveal more dispersions close to it by separating overlying cut-off frequencies. Extension of the measurement frequency range up to 10 MHz may also expose more dispersions. The correlation with structural changes in the cell matrix will lead to a complete model description of dielectric property changes during NET-formation. Furthermore, this model advances wound sensor development.

ACKNOWLEDGEMENTS

The authors would like to thank the Bundesministerium für Bildung und Forschung (BMBF) for funding the project "ChiBS—Chip-basierte Biologie für die Sensorik" (project number 45952) within the framework of the "WK-Potenzial" programme.

REFERENCES

[1] J. Dissemond, "Moderne Wundauflagen für die Therapie chronischer Wunden," *Der Hautarzt*, vol. 57, pp. 881–887, 2006.

[2] S. Wilm, A. Wollny, M. Rieger, and G. Gallenkemper, "Diagnostik und Therapie des Ulcus cruris venosum," tech. rep., Deutsche Gesellschaft für Phlebologie (DGP), 2010.

[3] P. Martin and S. Leibovich, "Inflammatory cells during wound repair: the good, the bad and the ugly," *Trends in Cell Biology*, vol. 15, no. 11, pp. 599–607, 2005.

[4] H. Schwan, "Electrical properties of tissue and cell suspensions," *Advances in Biological and Medical Physics*, vol. 5, pp. 147–209, 1957.

[5] K. Foster and H. Schwan, "Dielectric properties of tissues and biological materials: A critical review," *Critical Reviews in Biomedical Engineering*, vol. 17, pp. 25–104, 1989.

[6] M. Sakamoto, T. Fujikado, R. Hayakawa, and Y. Wada, "Low frequency dielectric relaxation and light scattering under AC electric field of DNA solutions," *Biophysical Chemistry*, vol. 11, no. 3, pp. 309–316, 1980.

[7] M. Schmeer, T. Seipp, U. Pliquett, S. Kakorin, and E. Neumann, "Mechanism for the conductivity changes caused by membrane electroporation of CHO cell-pellets," *Physical Chemistry Chemical Physics*, vol. 6, no. 24, pp. 5564–5574, 2004.

[8] F. Meuwly, P.-A. Ruffieux, A. Kadouri, and U. von Stockar, "Packed-bed bioreactors for mammalian cell culture: Bioprocess and biomedical applications," *Biotechnology Advances*, vol. 25, no. 1, pp. 45–56, 2007.

[9] R. Marquis and E. Carstensen, "Electric conductivity and internal osmolality of intact bacterial cells," *Journal of Bacteriology*, vol. 113, no. 3, pp. 1198–1206, 1973.

[10] T. Nacke, A. Barthel, D. Beckmann, J. Friedrich, M. Helbig, P. Peyerl, U. Pli-quett, and J. Sachs, "Messsystem für die impedanzspektroskopische breitband-prozessmesstechnik," *tm-Technisches Messen*, vol. 78, no. 1, pp. 3–14, 2011.

Electrode contacts for bioimpedance measurements

Lecture Notes on Impedance Spectroscopy, Volume 4 – Kanoun (ed)
© 2014 Taylor & Francis Group, London, ISBN 978-1-138-00140-4

Optimal electrode positions to determine bladder volume by bioimpedance spectroscopy

T. Schlebusch & S. Nienke
Philips Chair for Medical Information Technology (MedIT), RWTH Aachen University, Aachen, Germany

D. Leonhäuser & J. Grosse
Department of Urology, Urological Research, RWTH Aachen University Hospital, Aachen, Germany

S. Leonhardt
Philips Chair for Medical Information Technology (MedIT), RWTH Aachen University, Aachen, Germany

ABSTRACT: Continuous measurement of bladder volume is of great interest to patients with paraplegia. Since they lost the ability to percept bladder fullness and are unable to develop the urge to urinate, frequent self-catheterisation due to a fixed time scheme is necessary. Continuous, non-invasive measurement of bladder volume based on impedance measurements is a promising approach to advance from a fixed time scheme to a demand-driven self-catheterisation. In this work, the influence of electrode position on bladder volume measurement sensitivity is explored by finite element simulation.

Keywords: bioimpedance, bladder volume, electrode position

1 INTRODUCTION

1.1 *Motivation*

Especially for patients suffering from paraplegia, perception of urine accumulation as well as deliberate bladder emptying are impaired. The reasons are manifold, two examples are spinal cord injury or a damage of peripheral nervous structures, for example due to diabetes or dementia. In consequence, lifelong intermittent catheterisation has to be used by the patient for bladder emptying, as presented by [1]. In intermittent catheterisation, a sterile catheter is inserted into the bladder via the urethra, bypassing the sphincter muscle to drain the urine. Dependent on diuresis, this procedure has to be repeated several times a day; usually due to a fixed time scheme like every four hours. By choosing a short time window for urine accumulation, damages to the urinary tract by an over-distended bladder can be avoided. On the other hand, too frequent catheterisation is unnecessarily interfering the patient's life and increasing treatment costs by consuming more catheters.

A solution can be a demand-oriented catheterisation scheme, reducing the number of catheterisations to the necessary minimum. Therefore, a device measuring bladder volume unobtrusively and non-invasively would be of great benefit for patients with impaired reception of bladder pressure that are reliant on intermittent catheterisation. Such a device could for example be textile integrated or mounted to a wheelchair, keeping the patient informed about the current bladder volume and alarming him or her when emptying is necessary.

1.2 *Approaches for bladder volume estimation*

The technique most commonly used to measure urine volume in patients undergoing hospital examination like urodynamics is X-ray treatment, which is not suitable for continuous measurement. An alternate method is ultrasound imaging, which could be suitable for continuous or frequent use, but shows partially high error in calculated volume, even for skilled examiners [2, 3]. It is assumed that mainly the high variations in bladder shape are leading to erroneous volumes [4]. As an alternative, [5] showed a linear dependency of electrical impedance measurements and bladder volume. While they used electrodes attached directly to the bladder wall, [6] showed that the linear impedance trend is retained when measuring non-invasively by electrodes attached to the skin. Recently, this approach has been extended to an electrical impedance tomography approach to combine impedance and spatial information by [7]. One major difference between attaching the electrodes directly to the bladder wall or to the abdominal wall is that in the latter case attention has to be paid to the electrode positions to ensure that most of the current is flowing through the bladder region to maximize impedance variation with increasing bladder volume.

1.3 *Influences on abdominal impedance*

Unlike X-rays or ultrasound wave fronts, the electric current lines are highly dependent on the conductivity distribution in the volume itself. This implies that volume calculation from impedance measurements is not only affected by the urine volume, but also by impedance changes of surrounding tissue. In Figure 1 (a), five regions which affect the abdominal impedance measurement are shown. Although the skin and fat layer in region A are generally a static influence [8], movement of the fat layer when changing the body position might be an issue especially in adipose patients. When analysing current paths and comparing impedance data over a longer time period, impedance changes in the uterus (region D) have to be kept in mind for female subjects. Also, changes of intestine (region B) and colon (region E) impedance are influencing the measurements. In human and animal trials, the linear dependence of impedance and bladder volume has been proven, when all other influences are kept constant. To assess the influence of varying urine conductivity, a study has been done in freshly euthanised pigs. The results in Figure 2 show a clear influence of urine conductivity on abdominal impedance change per 50 ml filling volume. Therefore, the bladder in region C does not only affect the measurements by its varying filling volume, but also by varying urine conductivity.

In addition to changes in abdominal impedance itself, also the measurement setup can influence the results, for example by movement of electrodes or poor electrode-skin contact

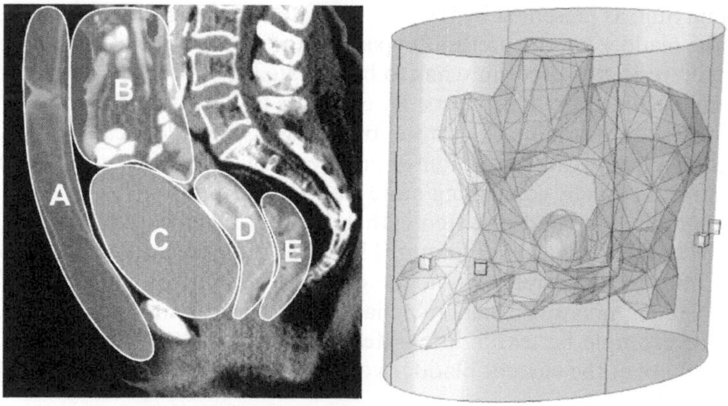
A
B
C
D
E

(a) Influences to abdominal impedance

(b) FEM Model

Figure 1. CT scan and simplified FEM model.

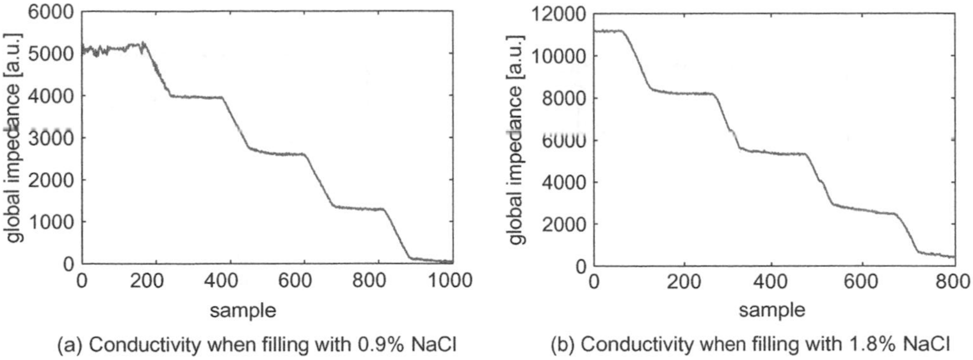
6000
5000
4000
3000
2000
1000
0
global impedance [a.u.]
0
200
400
600
800
1000
sample
(a) Conductivity when filling with 0.9% NaCl
12000
10000
8000
6000
4000
2000
0
global impedance [a.u.]
0
200
400
600
800
sample
(b) Conductivity when filling with 1.8% NaCl

Figure 2. Dependence of impedance on urine conductivity.

impedance. In this work, we would like to target the influence of electrode position and electrode movement on the volume sensitivity of an impedance-based bladder volume monitor. By using finite element simulations of the pelvic compartment, we will determine optimal electrode positions that maximize the variation of impedance per volume change.

2 MATERIALS AND METHODS

2.1 *Literature review on electrode positions for bladder impedance monitoring*

The first measurements by [6] used ring electrodes attached around the thorax, which measured the impedance of the whole abdomen from the diaphragm to the legs. A more bladder-specific approach has been used by [8] and [9], who attached two electrodes ventrally to the abdomen in the bladder region. To limit the influence of varying electrode-skin contact impedance, [7] used a four-electrode setup with the electrodes attached to the side on the abdomen. Alternatively, they also used an EIT-approach with an electrode belt of 16 electrodes. All these two or four electrode setups used electrodes arranged on a horizontal level around the abdomen. No set-ups using vertical alignments or ventral-to-dorsal alignments are known to the authors from literature. The only exception is the measurement by [10], who used a matrix attached to the abdomen ventrally. Therefore, three completely different electrode configurations have been addressed in this work, as presented in the next paragraph.

2.2 *FEM model for sensitivity analysis*

A finite element model of the pelvic region based on patient CT datasets has been build. The RWTH Aachen University hospital kindly supported DICOM datasets of patients with nearly empty and with greatly distended bladder, respectively. The iliac bone has been segmented, geometrically simplified and imported as a three dimensional volume to COMSOL Multiphysics 4.2a. Around the iliac bone, a cylinder with elliptic basic area has been placed to approximate the body shape. For the bladder, an ellipsoid distending from the floor of the pelvic cavity to cranial and ventral was included. The size of the bladder was parametrised, so that it is possible to simulate varying volumes. To analyse the effect of electrode position on measurement sensitivity, three arrangements were taken into consideration using the standard four point impedance measurement technique. For the first arrangement, all electrodes were placed ventrally on a horizontal line and then moved from caudal to cranial (Figure 1 (b) and Figure 3 (a)). For the second arrangement, all electrodes were placed ventrally on a vertical line and the upper electrode pair is moved to cranial (Figure 3 (b)). For the third case, one electrode pair was placed ventral, the counter electrode pair dorsal and all electrodes are shifted to cranial as in the first case (Figure 3 (c)). As tissue properties, cortical bone and

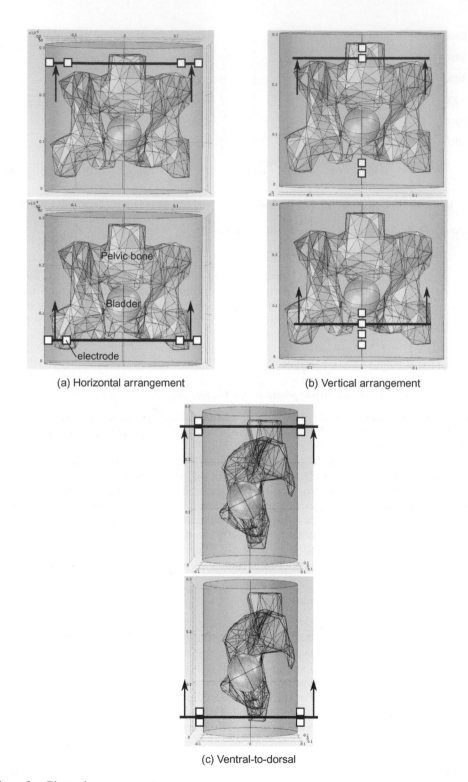

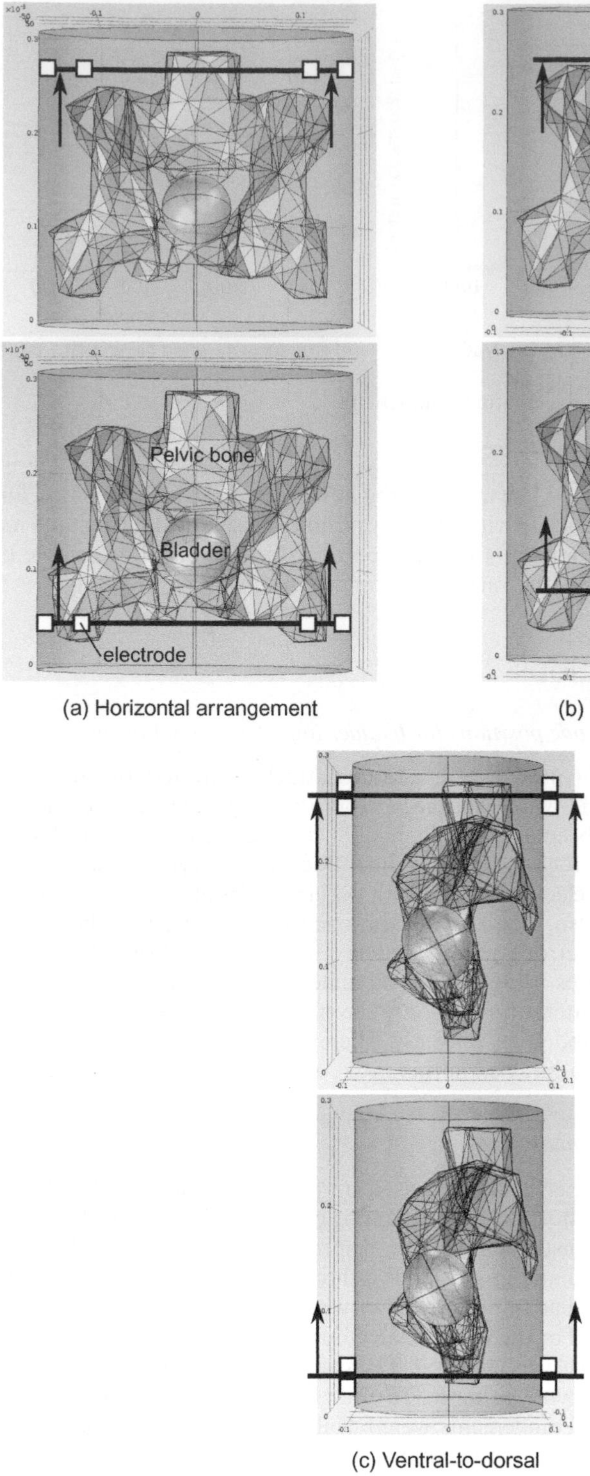
Pelvic bone
Bladder
electrode
(a) Horizontal arrangement
(b) Vertical arrangement
(c) Ventral-to-dorsal

Figure 3. Electrode arrangements.

colon were used from a COMSOL parameter file offered by [11] for download. The urine conductivity has been set to 30 mS/cm.

2.3 *Sensitivity in bioimpedance measurements*

For the simulation of an impedance measurement in COMSOL, one of the current injecting electrodes was assigned to a terminal with a unity current of 1 A. The second current electrode was assigned to ground. Boundary probes were assigned to the two voltage measurement electrodes and the differential voltage between them is being recorded. Since the current source was driving a unity current of 1 A, the resulting differential voltage is equal to the impedance between the voltage pick-up electrodes.

In addition to these general impedance simulations, the sensitivity of an electrode arrangement to impedance variations in the bladder region has been assessed. For this, the principle of the reciprocal lead field described by [12] and adapted to COMSOL by [11] has been used. To quantify to which extent a small volume contributes to the measured overall impedance, a second, virtual current source is introduced by making the voltage measurement electrodes current carrying. The current density vector field by this virtual current source is called the "reciprocal lead field" and specifies the sensitivity of the voltage measurement electrodes. The measured impedance can be expressed as

$$Z = \iiint \rho J_{inj}/I_{inj} \cdot J'_{reci} \, \mathrm{d}v \tag{1}$$

where Z is the measured impedance, ρ the conductivity of the sample, J_{inj} the current density field caused by by a current source of strength I_{inj} at the current injecting electrodes and J_{reci} is the reciprocal lead field from the virtual current source at the voltage measurement electrodes.

The sensitivity to the impedance of a small volume is maximized if the dot product of injected current density and reciprocal lead field is maximized. As the sensitivity field we denote

$$S = J'_{inj} \cdot J'_{reci} \tag{2}$$

Here the primes indicate that J'_{inj} and J'_{reci} are based on unity current. For the sensitivity assessment of the electrode configurations, the sensitivity field has been integrated in the bladder volume. A high integrated sensitivity expresses that most of the current is passing through the bladder volume and current injecting and voltage measurement electrodes are well aligned.

3 RESULTS

All simulations have been performed for four different bladder volumes (51, 102, 153 and 205 ml) and for several shift positions. The frequency has been fixed to 5 kHz.

The horizontal arrangement that is widely used in bladder impedance measurements (see sec. 2.1) shows relatively low sensitivity (see Figure 4 (a)). The highest sensitivity is reached when the electrodes are placed on the level of the bladder. When the electrodes are moved in cranial direction, the sensitivity decreases, which is in good agreement with our measurements. Figure 5 (a) shows the strong influence of electrode position on the resulting impedance measurement. Due to the steep slope even small electrode movements could cover volume based impedance changes. Figure 6 (a) shows the same impedance like Figure 5 (a), but the impedance for 51 ml is subtracted. It shows that the differential voltage per unit volume is relatively constant for all positions, so the high influence of the electrode position in Figure 5 (a) is only due to differences in static tissue impedance and not based on varying sensitivity.

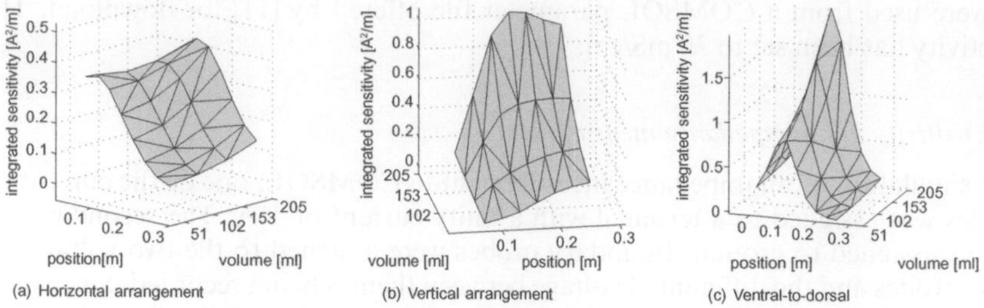

integrated sensitivity [A²/m]
position[m]
volume [ml]
(a) Horizontal arrangement
integrated sensitivity [A²/m]
volume [ml]
position [m]
(b) Vertical arrangement
integrated sensitivity [A²/m]
position [m]
volume [ml]
(c) Ventral-to-dorsal

Figure 4. Integrated sensitivity field depending on bladder volume and electrode position.

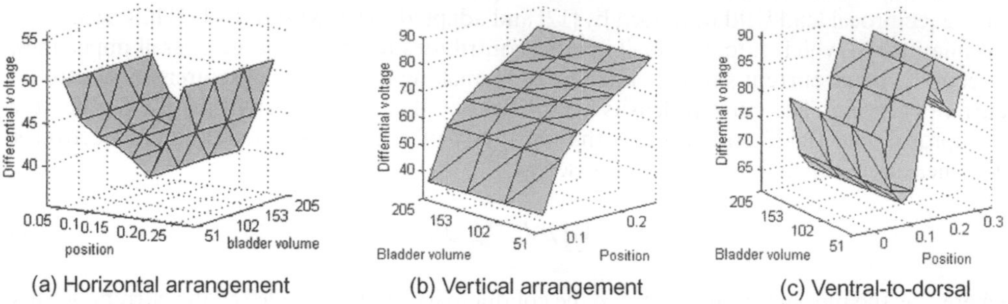

Differential voltage
position
bladder volume
(a) Horizontal arrangement
Differntial voltage
Bladder volume
Position
(b) Vertical arrangement
Differntial voltage
Bladder volume
Position
(c) Ventral-to-dorsal

Figure 5. Impedance depending on bladder volume and electrode position.

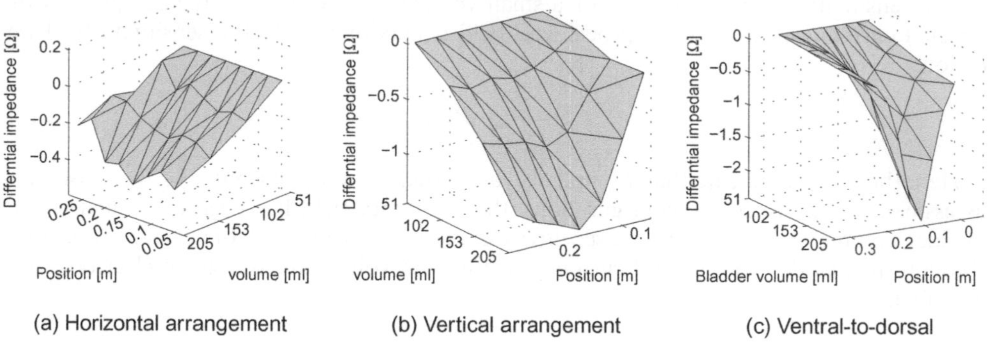

Differntial impedance [Ω]
Position [m]
volume [ml]
(a) Horizontal arrangement
Differntial impedance [Ω]
volume [ml]
Position [m]
(b) Vertical arrangement
Differntial impedance [Ω]
Bladder volume [ml]
Position [m]
(c) Ventral-to-dorsal

Figure 6. Differential impedance referenced to 51 ml.

The vertical alignment has good sensitivity over a wide range of positions for the upper electrode pair (see Figure 4 (b)). As expected the impedance in Figure 5 (a) is dependent on the distance of the pick-up electrodes and the slope of the differential impedance in Figure 6 (a) is relatively constant. If a minimum distance of around 15 cm between the electrode pairs is maintained, this configuration provides a good sensitivity with acceptable position dependence. The lower electrode pair then is located at the bladder base level and the upper pair above the maximum bladder extension.

The highest sensitivities are possible with the ventral-to-dorsal electrode alignment (see Figure 4 (c)). Unfortunately the impedance in Figure 5 (c) shows great dependence on electrode position and also the slope per unit volume is highly dependent on the position (see Figure 6 (c)). Highest sensitivity can only be reached if the bladder is exactly located between the electrode pairs and ultrasound imaging is highly suggested to assist during electrode placement.

4 CONCLUSION

Surprisingly, the most used horizontal alignment seems to have the poorest performance of the three configurations taken into account, at least from the simplified finite element simulations. Although the ventral-to-dorsal arrangement shows maximum sensitivity, it is very dependent on exact electrode placement, as illustrated by the steep surface in the results plot in Figure 6 (c). This could be a drawback in everyday use, so that the vertical arrangement states a good compromise of sensitivity and reproducibility in the simulations. Since the finite element model that has been used for the simulations is only a coarse approximation of the true anatomy, a patient study evaluating different electrode configurations is planned.

ACKNOWLEDGMENT

This work in the scope of the "UroWatch" project has kindly been supported by the German Federal Ministry of Education and Research (BMBF) within the research framework "innovative Hilfen" under funding number 01EZ1128A.

REFERENCES

[1] L. Guttmann and H. Frankel, "The value of intermittent catheterisation in the early management of traumatic paraplegia and tetraplegia," *Paraplegia*, vol. 4, no. 2, pp. 63–84, 1966.

[2] M. Dicuio, G. Pomara, F. Menchini Fabris, V. Ales, C. Dahlstrand, and G. Morelli, "Measurements of urinary bladder volume: Comparison of five ultrasound calculation methods in volunteers," *Arch Ital Urol Androl*, vol. 77, no. 1, pp. 60–62, 2005.

[3] S. Hynds, C.K. McGarry, D. Mitchell, S. Early, L. Shum, D. Stewart, J.A. Harney, C. Cardwell, and J. O'Sullivan, "Assessing the daily consistency of bladder filling using an ultrasonic bladderscan device in men receiving radical conformal radiotherapy for prostate cancer," *British Journal of Radiology*, vol. 84, no. 1005, pp. 813–818, 2011.

[4] C. Nwosu, K. Khan, P. Chien, and M. Honest, "Is real-time ultrasonic bladder volume estimation reliable and valid? A systematic overview," *Scandinavian Journal of Urology and Nephrology*, vol. 32, no. 5, pp. 325–330, 1998.

[5] M. Talibi, R. Drolet, H. Kunov, and C. Robson, "A model for studying the electrical stimulation of the urinary bladder of dogs1," *British Journal of Urology*, vol. 42, no. 1, pp. 56–65, 1970.

[6] J. Denniston and L. Baker, "Measurement of urinary bladder emptying using electrical impedance," *Medical and Biological Engineering and Computing*, vol. 13, no. 2, pp. 305–306, 1975.

[7] S. Leonhardt, A. Cordes, H. Plewa, R. Pikkemaat, I. Soljanik, K. Moehring, H. Gerner, and R. Rupp, "Electric impedance tomography for monitoring volume and size of the urinary bladder," *Biomedizinische Technik/Biomedical Engineering*, vol. 56, no. 6, pp. 301–307, 2011.

[8] W. Liao and F. Jaw, "Noninvasive electrical impedance analysis to measure human urinary bladder volume," *Journal of Obstetrics and Gynaecology Research*, vol. 37, pp. 1071–1075, 2011.

[9] C. Kim, T. Linsenmeyer, H. Kim, and H. Yoon, "Bladder volume measurement with electrical impedance analysis in spinal cord-injured patients," *American Journal of Physical Medicine & Rehabilitation*, vol. 77, no. 6, pp. 498–502, 1998.

[10] P. Hua, E. Woo, J. Webster, and W. Tompkins, "Bladder fullness detection using multiple electrodes," in *Engineering in Medicine and Biology Society, 1988. Proceedings of the Annual International Conference of the IEEE*, pp. 290–291, IEEE, 1988.

[11] F. Pettersen and J. Høgetveit, "From 3d tissue data to impedance using simpleware scanfe+ ip and comsol multiphysics–a tutorial," *Journal of Electrical Bioimpedance*, vol. 2, no. 1, pp. 13–32, 2011.

[12] O. Martinsen and S. Grimnes, *Bioimpedance and Bioelectricity Basics*. Academic press, 2008.

Lecture Notes on Impedance Spectroscopy, Volume 4 – Kanoun (ed)
© 2014 Taylor & Francis Group, London, ISBN 978-1-138-00140-4

Theoretical and experimental comparison of microelectrode sensing configurations for impedimetric cell monitoring

M. Carminati
Dipartimento di Elettronica e Informazione, Politecnico di Milano, Milano, Italy

C. Caviglia & A. Heiskanen
Department of Micro-and Nano-Technology, Technical University of Denmark, Lyngby, Denmark

M. Vergani, G. Ferrari & M. Sampietro
Dipartimento di Elettronica e Informazione, Politecnico di Milano, Milano, Italy

T.L. Andresen & J. Emnéus
Department of Micro-and Nano-Technology, Technical University of Denmark, Lyngby, Denmark

ABSTRACT: A theoretical and experimental comparison between vertical and coplanar interdigitated sensing configurations for impedimetric cell growth tracking is presented. These widely-adopted approaches are quantitatively compared on the same cell populations and on the same 10 μm interdigitated microelectrodes using a versatile custom-made monitoring platform including a 24-channel miniaturized potentiostat. The characterization of bare microelectrodes in buffer and tracking experiments with HeLa cells over 16 hours demonstrate that the coplanar configuration provides a higher sensitivity to cell adhesion and spreading (Cell Index = 1.6 vs. 0.4) albeit at a higher frequency of maximum sensitivity (100 kHz vs. 24 kHz) shifting over time.

Keywords: impedance spectroscopy, interdigitated electrodes, cell tracking

1 INTRODUCTION

The use of impedance spectroscopy, consolidated in various electrochemical applications such as corrosion studies or characterization of batteries, is becoming a routine tool also in the biomedical field, in particular for monitoring the growth and response to chemical stimulation of cell populations cultivated on planar microelectrodes. In fact, since the pioneering work by Giaever and Keese [1], a huge volume of research has flourished and various custom-made and commercial ECIS (Electric Cell-substrate Impedance Sensing) systems are available for biological investigation [2]. The success of this technique, based on the increase of impedance due to the insulating barrier represented by the cells filling the volume above the electrodes, relies on its simplicity enabling label-free and automatic analysis. However, as highlighted by Orazem and Tribollet [3], as impedance is not specific by itself and is affected by any variation of the interface properties, the interpretation of impedance data is particularly delicate. Elaborate equivalent models, though providing good fitting, can lead to the loss of the physical meaning of their parameters.

In this work we present a theoretical and experimental comparison of two alternative sensing configurations, i) the standard "vertical" configuration (a single working electrode versus a large counter electrode, and ii) the interdigitated configuration (one comb versus the other one closely spaced). The latter was introduced for cellular monitoring 15 years ago [4] and is now becoming commonly adopted due to its good performance [5] even in commercial systems, such as exCELLigence by Roche. The major novelty of our investigation is that for the first time a quantitative comparison is performed exactly on the same cells, alternating the two configurations when measuring impedance spectra on the same couples of electrodes, as enabled by an original and versatile monitoring platform that we have recently designed [6].

2 THEORY AND MODELING

2.1 *Quantitative interface modeling*

Differently from other kinds of impedimetric biosensor based on the modulation of the interfacial properties, in this case all the measurements are performed in standard buffer solutions (PBS—*Phosphate Buffered Saline* o cell culture medium that has similar properties), without the addition of any redox species. This allows avoiding toxicity issues, possible electrochemical interferences and preserves the simplicity and generality of the proposed technique. Thus, in the absence of Faradaic processes, no charge transfer mechanism takes place and the interfacial impedance is simply formed by the series of the double layer capacitance and the solution resistance. For the gold/PBS interface, the double layer capacitance can be estimated multiplying the area of the electrode exposed to the liquid (down to the micro-scale) by the specific interfacial capacitance being 0.1–0.2 pF/μm^2.

A huge amount of experimental data has shown that the real behavior of the metal/solution interface is far from an ideal capacitance and that usually measured spectra are better described by constant-phase elements (CPE). A CPE is characterized by a pseudo capacitive impedance $1/(2\pi \cdot i \cdot f \cdot Q)^n$ with $n < 1$ (i.e. a magnitude slope smaller than –20 dB/dec and a phase of –90° · n). Although a unified theory relating the CPE parameters (mainly n, typically ranging from 0.7 to 0.9) to the physical properties of a blocking electrode is still missing (n can be related to the fractal dimension of the surface [7]), it is commonly accepted that a CPE is originated by a spatial 2D or 3D non uniform distribution of time constants [3, 8], mainly due to the roughness and porosity of the electrode surface. It results that the higher is the roughness, the smaller is n. Thus, as the used electrode are very smooth (measured $n = 0.9$) in the following we will treat the interface as an ideal C_{DL} capacitance with no impact on the design considerations nor on the results interpretation.

On the contrary, the solution resistance is a bulk property, set by the mobility and by the concentration of the ions in the electrolytic solution. For PBS the resistivity ρ is 0.66 Ω. It must be taken into account that at the micro-scale this resistance no longer scales with the inverse of the area of the electrode but with its perimeter. For instance, for a planar disk electrode it is inversely proportional to the radius r and can be estimated as $R_{SOL} = \frac{\rho}{4r}$ [9].

2.2 *Vertical vs. coplanar setting*

The two alternative configurations that we compare are illustrated in Fig. 1a. In the "vertical" case, the small sinusoidal excitation voltage (typically in the mV range to avoid non-linearity and to limit the perturbation applied to the cells) is applied to a large and distant Counter Electrode (CE) while the current is read at the Working Electrode (WE). As usual in standard electrochemical setups, as the CE is significantly larger than the WE, its interfacial impedance can be neglected and the lumped-parameter small-signal equivalent impedance model corresponds to the series of the WE C_{DL} and R_{SOL1}, as shown in Fig. 2b. The WE, fabricated in gold on an oxidized silicon substrate with a standard lift-off-process [6] and passivated with Si_3N_4 layer which was reactive ion etched to expose the active area of 6×10^4 μm^2 being formed by 12 fingers of 10 μm width and 500 μm length. This size has been chosen in order to have a total sensing area comparable with standard ECIS commercial electrodes (such as the 8 W2 × 1E array of 16 250 μm disks by Applied Biophysics). The expected value of C_{DL} is thus ≈ 6 nF. The custom multichannel potentiostat used in these experiments [6] has been designed to cope with such a high input capacitance, critical for the bandwidth and stability of the input circuits and for the current noise performance of the potentiostat. The value of R_{SOL1} can be estimated calculating the equivalent radius (138 μm) of a disk electrode with the same area, giving 1.2 kΩ.

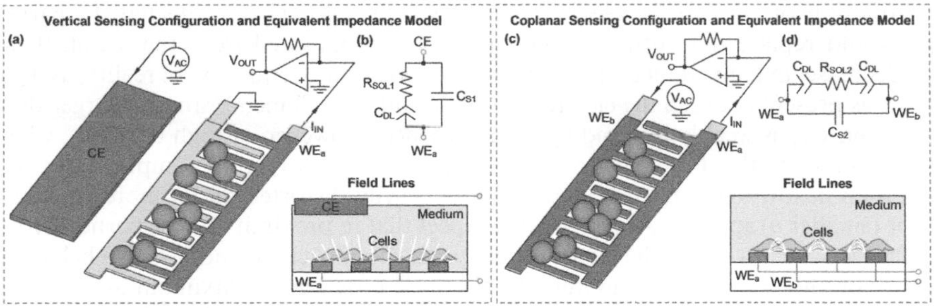
Vertical Sensing Configuration and Equivalent Impedance Model
(a)
(b)
CE
V_AC
V_OUT
CE
R_SOL1
C_S1
C_DL
I_IN
WE_a
WE_a
Field Lines
CE
Medium
Cells
WE_a
Coplanar Sensing Configuration and Equivalent Impedance Model
(c)
(d)
V_OUT
C_DL R_SOL2 C_DL
V_AC
WE_b
I_IN
WE_a
WE_b
C_S2
WE_a
Field Lines
Medium
Cells
WE_a WE_b

Figure 1. Compared vertical and coplanar sensing configurations, cross-section cartoons showing the field lines and equivalent impedance models.

The coplanar configuration is illustrated in Fig. 1c. In this case the current is measured at the same electrode WEa, but the sinusoidal stimulation is applied to the companion comb of electrode fingers (WEb), interdigitated with WEa at a distance of 10 μm. As the vertical penetration of the electric field between interdigitated electrodes is roughly equal to the gap, this size has been chosen in order to probe cells in the 5 μm–20 μm size range. In this case the two interfaces are exactly symmetrical and the corresponding impedance model, shown in Fig. 1d, contains the two C_{DL}'s bridged by the solution resistance R_{SOL2} in the middle. As the two C_{DL}'s are in series, the resulting total capacitance will be $C_{DL}/2$. Instead, given the particular geometry (dominated by fringing effects), the estimation of the solution resistance of the volume above the fingers is more complicated. Approximated but very useful analytical expressions are obtained by means of conformal mapping, i.e. a set of mathematical transformations used to map the coplanar geometry to the parallel-plate case for the calculation of the cell constant k, relating the resistance R and capacitance C between the combs to the resistivity ρ and permittivity ε_r of the material above the electrodes respectively: $R = k \cdot \rho$ and $C = \varepsilon_0 \cdot \frac{\varepsilon_r}{k}$. even for the multi-layer heterogeneous case [10]. For our geometry in PBS, the expected R_{SOL2} is $\approx 100\ \Omega$.

In both cases, a physical capacitance C_S must be added in parallel to the models accounting for the total amount of stray capacitance present between the electrodes metals. This stray capacitance is the sum of several contributions: The coupling between the connection wires, the coupling of the metal tracks through the substrate and the coupling of the passivated tracks running beneath the solution. This capacitance should be minimized during the design of the electrodes as it can produce a significant degradation of the accuracy and resolution of the impedance measurement [11].

2.3 *Sensitivity to cell proliferation*

Tracking over time adhesion, proliferation, response to chemo-mechanical stimuli and death of a colony of adherent cells cultured on top of these planar electrodes is commonly performed at one single frequency (or a few discrete ones). Thus, the purpose of the analysis of the impedance over a wide frequency range is to identify the spectral range where the sensitivity of impedance to cell coverage is the highest. This sensitivity is affected by various delicate aspects, including cell morphology, adhesion levels, functionalization of electrode surface etc… Correspondingly, in the literature several equivalent models have been proposed to describe the variation of the impedance produced by the cell coverage. The most common one consists in the insertion of an intermediate $R_{cell} \| C_{cell}$ parallel couple between C_{DL} and R_{SOL} accounting for the dielectric (C_{cell}) and dissipative (R_{cell}) properties of the layer of cells [12]. Although very general, we have chosen to further simplify this model and adopt the same models of Fig. 1 even in the presence of cells, considering the variations of both parameters induced by cell adhesion. As cells proliferate, the impedance magnitude increases, mainly because at frequencies below ≈10 MHz cells behave like an insulating barrier and the conductive paths of ions are obstructed by them, thus increasing R_{SOL}. In principle, if the adhesion of cells to the electrodes is very tight, also C_{DL} is affected. In particular, if cells would replace the diffused layer and ions are separated from the metal, then C_{DL} should decrease and impedance should increases also at low frequency. In reality, as a gap of 10–20 nm is present between the electrode surface and the cell membrane, i.e. larger than the double layer, C_{DL} is weakly affected by their presence. However, at high densities, when the access resistance to the double layer of ions "percolating" through a compact layer of cells is high, even at low frequency the impedance is heavily distorted and a more evident CPE behavior (smaller n) appears. As a result, we expect that in proximity of the corner frequency $f_c = 1/(2\pi \cdot R_{SOL} \cdot C_{DL})$, i.e. at the transition between the capacitive and resistive behavior, the sensitivity of the impedance magnitude to electrode coverage is maximum as both parameters are changing. This has been confirmed by different experimental results reported in the literature [6,12] in which a peaked sensitivity is observed around f_c.

On the basis of the quantitative analysis illustrated in Sec. III.B, we can expect that when passing from the vertical to the coplanar configuration, the measured impedance spectra will change significantly: At low frequency (where C_{DL} dominates) impedance is doubled, while at high frequency (where R_{SOL} dominates) the resistive plateau is lower. Thus, the corner frequency will increase for both a reduction of R_{SOL} and C_{DL}. We consequently conclude that for the vertical configuration the optimal operating frequency for cell growth tracking (i.e. f_c) is lower than in the coplanar case.

On the other hand, for small initial cell densities (i.e. small fractions of covered electrode area, far from confluence) it is expected that the coplanar detection is more sensitive to the presence of cells in the volume between the fingers, with respect to the vertical condition in which R_{SOL} is set by the electrode perimeter. At frequencies $\geq f_c$, the electric field intensity is more confined in the volume occupied by the cells in the coplanar geometry, in analogy to several successful examples of high speed cell detection in microfluidic flow cytometry [13].

In conclusion, the coplanar configuration is expected to provide higher absolute sensitivity to cell proliferation despite a higher optimal operating frequency. The following experiments are carried out in order to verify this hypothesis and investigate the frequency/sensitivity trade-off on a quantitative basis.

3 MATERIALS AND METHODS

The monitoring platform is composed by two main elements: A cell culture device integrated with a microelectrode chip and a multichannel bipotentiostat [6]. The cell culture device consists of a micromilled poly (methyl methacrylate) (PMMA) holder characterized by an upper vial, containing up to 600 µl of volume, that allows cell seeding and culture medium exchange.

HeLa cells and 3T3 Fibroblasts were routinely grown in 75 cm^2 cell culture flasks at 37 °C with 5% CO_2 in a humidified atmosphere. Dulbecco's Modified Eagle's medium supplemented with 10% fetal bovine serum and 1% penicillin/streptomycin was used as culture medium for HeLa cell line and fibroblasts. After reaching the 80–90% confluence, the cells were washed with Dulbecco's Phosphate Buffered Saline (PBS) and harvested through trypsinization.

Electrochemical Impedance Spectroscopic (EIS) tracking has been used as a non-invasive biophysical technique to continuously monitor and compare cell adhesion and proliferation of HeLa cells. Prior to using, in order to clean the gold surface of the microelectrode chip from the ambient contaminants, the chips have been treated with a 25% H_2O_2/50 mM KOH solution for 10 minutes. The culture chamber has been sterilized using 500 mM NaOH and subsequently and to facilitate cellular adhesion, the electrode chip has been coated with poly-L-lysine (10 µg/mL). About 2.5 $\times 10^5$ cells have been initially seeded into the chamber and the measurements were performed while the monitoring platform was incubated at 37°C, in a humidified atmosphere with 5% CO_2. Impedance spectra were acquired from each sensor element for 16 hours: During the first 4 hours after cell seeding data were recorded from each sensor element every 20 minutes, then every hour. In order to limit the current flowing through the cells, a 200 µV sinusoidal perturbation potential was applied (to reduce the max.

current below the 1 μA safety limit) and 30 points were recorded in the frequency range between 100 Hz and 100 kHz, each with an averaging time of 2s. All calculations and graphs have been obtained using MATLAB® software. For each experiment, data acquired from different sensor elements were averaged and processed to derive a parameter termed Cell Index (*CI*) that represents the relative impedance change in the system. For each frequency and for each time point the *CI* is calculated as $CI(t, f) = (|Z(t, f)ght|/|Z(0, f)|) - 1$, where $|Z(0, f)|$ is the magnitude of impedance acquired at the start of the experiment, when cells were not yet seeded into the well and $|Z(t, f)|$ is the magnitude of impedance acquired after cell seeding at different time points. Data are represented as mean ± standard deviation.

4 EXPERIMENTAL RESULTS

4.1 *Electrodes characterization*

Prior to the experiments with cell populations, the interfacial impedance of the bare electrodes in contact with PBS has been characterized. A commercial LCR meter (Agilent E4980a) has been used instead of the custom potentiostat (limited to 100 kHz due to USB data transfer constraints) in order to explore a higher frequency range and probe the resistive plates in both cases. The measured spectra (continuous lines in Fig. 2) excellently match the theoretical values (dashed lines), confirming the correct estimates of C_{DL} of WEa in both cases (i.e. vs. CE and WEb respectively). The value of the measured capacitance matches the expected double layer value (from theory we expect for PBS a specific capacitance of 0.1–0.2 pF/μm^2) and the slightly sub-capacitive slope is fitted by a CPE exponent = 0.9. Given the proximity of the exponent to 1 (thanks to a good smoothness of the surface of the micro fabricated electrodes), it is treated in the following as an ideal capacitor. Despite the large area and the use of the expression for the disk electrodes, R_{SOL1} has been very precisely estimated. The measured R_{SOL2} is twice the value calculated with conformal mapping, but however 6 times smaller than in the vertical case.

As expected, the corner frequency shifts from ≈ 30 kHz (vertical) to ≈ 300 kHz (coplanar). The effect of the stray capacitance is visible in the measured spectrum for the vertical configuration where a pole at 1 MHz is visible (corresponding to a C_{S1} = 130 pF, due to large couplings through the silicon substrate and to the long metal tracks below the passivation, but still negligible with respect to C_{DL} = 6 nF).

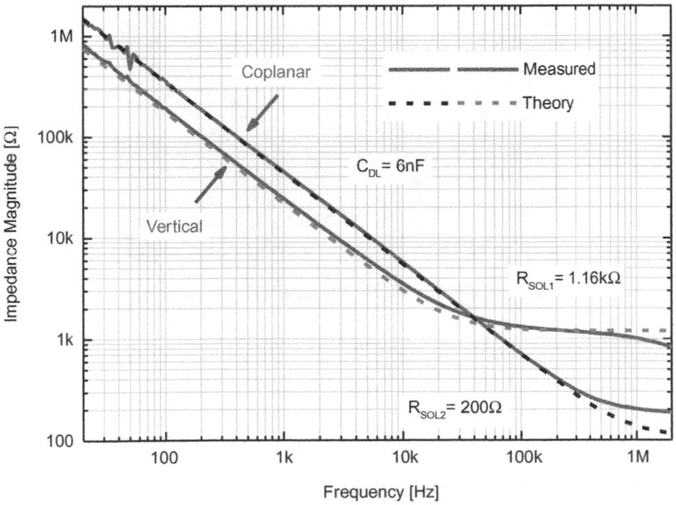

Coplanar
Measured
Theory
C_{DL} = 6nF
Vertical
R_{SOL1} = 1.16kΩ
R_{SOL2} = 200Ω
Impedance Magnitude [Ω]
1M
100k
10k
1k
100
100
1k
10k
100k
1M
Frequency [Hz]

Figure 2. Compared theoretical and measured spectra in PBS for both configurations (V_{AC} = 4 mV) with Agilent LCR meter E4980a.

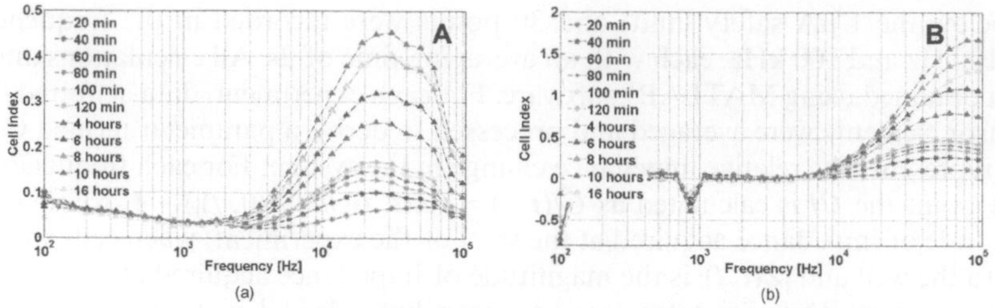
A
B
20 min
40 min
60 min
80 min
100 min
120 min
4 hours
6 hours
8 hours
10 hours
16 hours
Cell Index
Frequency [Hz]
(a)
(b)

Figure 3. Cell Index over frequency at different time instants after cells seeding: (a) vertical configuration peak at 24 kHz and (b) coplanar configuration, peak at 100 kHz shifting over time towards higher frequencies.

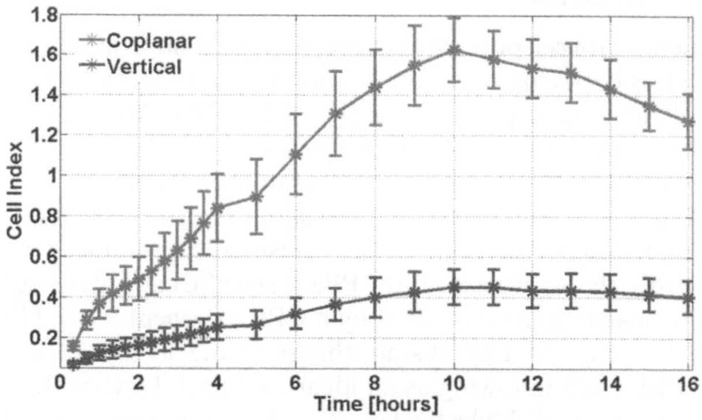
Coplanar
Vertical
Cell Index
Time [hours]

Figure 4. Cell Index tracking over time at 24 kHz and 100 kHz respectively for vertical and coplanar configuration. V_{AC} is reduced to 200 μV to limit the electromagnetic perturbation stress applied to the cell culture.

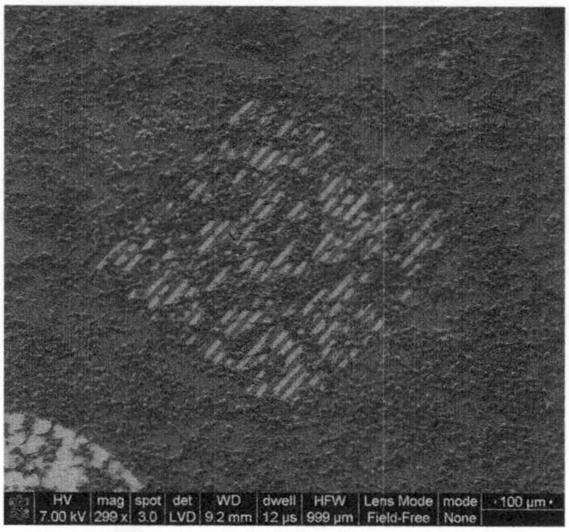
HV | mag | spot | det | WD | dwell | HFW | Lens Mode | mode | 100 μm
7.00 kV | 299 x | 3.0 | LVD | 9.2 mm | 12 μs | 999 μm | Field-Free | None

Figure 5. SEM images of HeLa cells on the WE interdigitated electrodes 24 hours after cell seeding (scale bar is 100 μm) confirming good adhesion and moderate density.

4.2 *Monitoring HeLa cells*

In order to evaluate the behavior of the detection platform and compare the two alternative detection configurations, electrochemical impedance spectroscopic monitoring of HeLa cells has been performed. Fig. 3 shows examples of *CI* values acquired using the ECIS vertical configuration. As expected, a peak frequency, indicating the most sensitive region of the spectra, can be determined at about 24 kHz for the vertical case, in excellent agreement with the 30 kHz obtained from the spectra of Fig. 2. Data recorded using the interdigitated coplanar configuration showed instead a peak frequency at higher frequency, again as expected. Fig. 4 shows the *CI* tracking, over 16 hours, acquired using the two different configuration modes and plotted at the characteristic peak frequency. The maximum value of *CI*, about $1.6 \pm 16\%$ for the coplanar and $0.45 \pm 17\%$ for the vertical configuration, was reached in both cases 10 hours after cell seeding. This clearly demonstrates that the coplanar configuration provides the highest sensitivity for monitoring cell adhesion and proliferation in particular at low and moderate seeding densities (see Fig. 5).

A reduction is *CI* for the coplanar configuration is observed after 10 hours and this is due to shifting of the peak to higher frequencies with respect to the fixed detection frequency (100 kHz). This shifting is due to a reduction of R_{SOL2} probably caused by a slight medium evaporation.

5 CONCLUSIONS

The presented electrochemical impedance monitoring platform allows parallel measurements to study biological phenomena. In order to optimize cell proliferation assays, two alternative configuration modes have been experimentally compared in parallel using HeLa cells as model cell line. It has been clearly proved that for monitoring cell adhesion, spreading and proliferation the coplanar configuration provides the highest sensitivity. Furthermore, results showed that as during long-term experiments, the solution resistance changes (typically decreasing, due for instance to slow medium evaporation or cell metabolism) correspondingly the position of the peak of maximum sensitivity shifts. Thus, instead of single-frequency recording, it is advisable to record a wider frequency span in order to be able to adaptively track the frequency of maximum sensitivity, avoiding artifacts. Based on these results, further developments will bring the system towards the implementation of a microfluidic platform that will allow long term parallel measurements solving the problem of the medium evaporation.

ACKNOWLEDGEMENTS

Financial support is acknowledged from EU through FP7 project EXCELL and from Fondazione CARIPLO.

REFERENCES

[1] I. Giaever and C. Keese, "Monitoring fibroblast behavior in tissue culture with an applied electric field," *Proceedings of the National Academy of Sciences*, vol. 81, no. 12, pp. 3761–3764, 1984.

[2] J. Hong, K. Kandasamy, M. Marimuthu, C. Choi, and S. Kim, "Electrical cell-substrate impedance sensing as a non-invasive tool for cancer cell study," *Analyst*, vol. 136, no. 2, pp. 237–245, 2011.

[3] M. Orazem and B. Tribollet, *Electrochemical Impedance Spectroscopy*, vol. 48. Wiley-Interscience, 2011.

[4] R. Ehret, W. Baumann, M. Brischwein, A. Schwinde, K. Stegbauer, and B. Wolf, "Monitoring of cellular behaviour by impedance measurements on interdigitated electrode structures," *Biosensors and Bioelectronics*, vol. 12, no. 1, pp. 29–41, 1997.

[5] J. Mamouni and L. Yang, "Interdigitated microelectrode-based microchip for electrical impedance spectroscopic study of oral cancer cells," *Biomedical Microdevices*, vol. 13, no. 6, pp. 1075–1088, 2011.

[6] M. Vergani, M. Carminati, G. Ferrari, E. Landini, C. Caviglia, A. Heiskanen, C. Comminges, K. Zor, D. Sabourin, M. Dufva, *et al.*, "Multichannel bipotentiostat integrated with a microfluidic platform for electrochemical real-time monitoring of cell cultures," *IEEE Transactions on Biomedical Circuits and Systems*, vol. 6, pp. 498–507, 2012.

[7] L. Nyikos and T. Pajkossy, "Fractal dimension and fractional power frequency-dependent impedance of blocking electrodes," *Electrochimica Acta*, vol. 30, no. 11, pp. 1533–1540, 1985.
[8] B. Hirschorn, M.E. Orazem, B. Tribollet, V. Vivier, I. Frateur, and M. Musiani, "Determination of effective capacitance and film thickness from constant-phase-element parameters," *Electrochimica Acta*, vol. 55, no. 21, pp. 6218–6227, 2010.
[9] J. Newman, *Electrochemical Systems*. Prentice-Hall, Inc., Englewood Cliffs, New Jersey, 1973.
[10] R. Igreja and C. Dias, "Analytical evaluation of the interdigital electrodes capacitance for a multi-layered structure," *Sensors and Actuators A: Physical*, vol. 112, no. 2, pp. 291–301, 2004.
[11] M. Carminati, M. Vergani, G. Ferrari, L. Caranzi, M. Caironi, and M. Sampietro, "Accuracy and resolution limits in quartz and silicon substrates with microelectrodes for electrochemical biosensors," *Sensors and Actuators B: Chemical*, vol. 174, pp. 168–175, 2012.
[12] L. Wang, H. Yin, W. Xing, Z. Yu, M. Guo, and J. Cheng, "Real-time, label-free monitoring of the cell cycle with a cellular impedance sensing chip," *Biosensors and Bioelectronics*, vol. 25, no. 5, pp. 990–995, 2010.
[13] T. Sun and H. Morgan, "Single-cell microfluidic impedance cytometry: A review," *Microfluidics and Nanofluidics*, vol. 8, no. 4, pp. 423–443, 2010.

Lecture Notes on Impedance Spectroscopy, Volume 4 – Kanoun (ed)
© 2014 Taylor & Francis Group, London, ISBN 978-1-138-00140-4

Influence of a sliding contact on impedance monitoring in a smart surgical milling tool

C. Brendle
Philips Chair for Medical Information Technology, RWTH Aachen University, Aachen, Germany

A. Niesche, A. Korff & K. Radermacher
Chair of Medical Engineering, RWTH Aachen University, Aachen, Germany

S. Leonhardt
Philips Chair for Medical Information Technology, RWTH Aachen University, Aachen, Germany

ABSTRACT: The risk for the patient during revision total hip replacement surgery is high. The present-day surgery standard procedures for the bone cement removal are manually controlled by using hammer and chisel, which may result in severe complications. To increase safety for the patient, we are developing an Impedance Controlled Surgical System, which should enable real-time controlled bone cement removal. For this purpose, we want to use the milling head, as measurement electrode during the milling process. To close the impedance measurement circuit over the milling head we have to transfer the alternating current on the fast rotating milling shaft. Therefore we need to identify the influence of a Sliding Contact System by measuring different test impedances in the rotating part of the system. We show the functionality of the sliding contact, characterize the capacitive coupling and quantity the noise caused by the rotation. Finally, these parameters are discussed in the context of complete impedance measurement chain.

Keywords: Bioimpedance spectroscopy, Surgical Instrumentation, Artificial Hip Replacement, Bone Cement milling, Sliding Contact

1 INTRODUCTION

The Revision of Total Hip Replacement (RTHR) is a medical standard procedure with an increasing number of cases (e.g.> 22.500 cases in Germany in 2008) [1]. Furthermore, the major part of the implanted Artificial Hip Joints (AHJ) is cemented and 5% to 12% of them are replaced after ten years [2, 3]. To ensure the fixation of the new implant, the old Bone Cement (BC) has to be completely removed [4]. Conventionally, this is manually done by using hammer and chisel. Besides lateral femur windowing may be used to resolve the restricted access to the operation field in the femur cavity. These procedures are highly invasive and complications (e.g. femur cracking, vessel traumata) appear [5].

Novel surgery strategies with navigated milling systems enable conservative BC removal, where the registration of the spatial BC is the major challenge. These upcoming approaches use ultrasound or Computed Tomography (CT) measurements to gain the information in an additional preparation step [2, 6, 7]. This provides additional stress for the patient (e.g. operation time, radiation) and unobserved intraoperative navigation offsets cannot be recognized.

To avoid these disadvantages, a real time milling process control has to be established, which we intent with the Impedance Controlled Surgical Instrumentation (ICOS) (Fig. 1). Due to the fact that the milling tool covers the cutting area, the measurement system has to be non-visually. For this purpose, we measure the electric impedance between the milling head and the

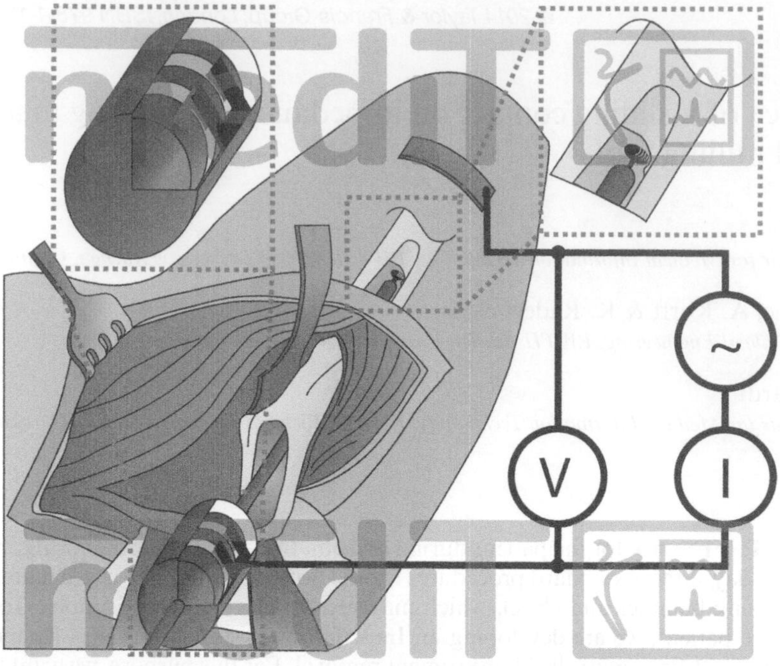
V
I
~

Figure 1. Operation scenario with the ICOS System.

human leg skin. In previous studies, we could approximate the wall capacity of a holy cylinder with a wall thickness of 1 mm out of Polymethylmethacrylate (PMMA) as BC substitute to $C_{PMMA} = 0.23$ pF. Due to the fact that the magnitude of this capacity is several orders higher than the magnitude of tissue impedance [8], we expect that the impedance signal is characterized by the residual BC thickness. To apply the alternating measurement current on to the rotating shaft, we explore a Sliding Contact System (SCS). This means that the SCS is a central part of the measurement chain. For this reason its characteristics have to be analyzed.

2 METHODS

The experimental setup (see Fig. 2) is driven by a motor system (Maxon Motor AG, Switzerland). The motor is coupled with an isolated shaft which passes the SCS. The SCS has two isolated hardened gold sliding contacts for current transfer between the outside stationary and the inside rotating part of the system. On the SCS jacket is the LCR-meter terminal for connecting the Agilent E4890A LCR-Meter to the sliding contacts, whose measurement signal is recorded with MathWorks® MATLAB. Next to the shaft in the rotating part of the SCS on the motor averted side are two isolated single conductors, which are used to connect the test impedances $R_i \in [10, 10.000\ k\Omega]$. During the experiments, we sweep the measurement frequency f_j between 1 and 2.000 kHz in 20 logarithmic scaled steps, we set the rotational speed ω between 0 and 8.000 1/min and we record 3 repetitons with 100 samples for each parameter combination. The cutoff frequency $f_{c,i}$ for each R_i is defined in the maximum measured frequency, where the deviation of the measured impedance mean value $\mu_{|Z(f_j,R_i,\omega_l)|}$ of the ideal value R_i is ≤ 5% for all ω_l:

$$f_{c,i} = max\left\{f_j \mid \mu_{|Z(f_j,R_i,\omega_l)|} \geq 0{,}95 \cdot R_i\right\} \tag{1}$$

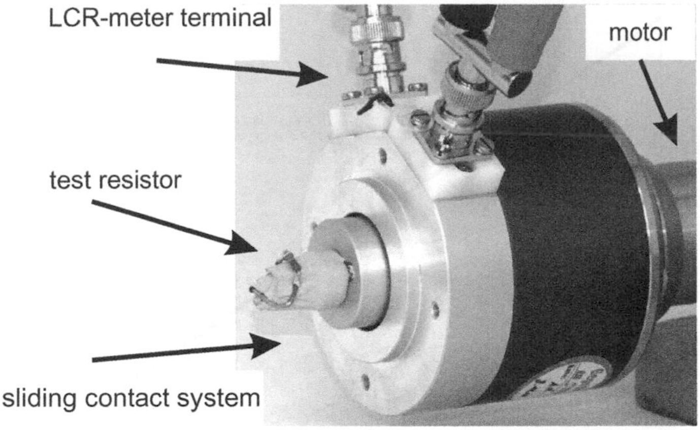
LCR-meter terminal
motor
test resistor
sliding contact system

Figure 2. Experimental setup of the SCS.

For characterizing the noise caused by the rotating SCS, we defined the Signal to Noise Ratio (SNR) in eqn. (2) in dependency of the mean $\mu_{|Z(f_j,R_i,\omega_l)|}$ and the standard deviation $\sigma_{|Z(f_j,R_i,\omega_l)|}$ of the absolute impedance value $|Z(f_j, R_i, \omega_l)|$.

$$SNR = \log_{10} \frac{\mu_{|Z(f_j,R_i,\omega_l)|}}{\sigma_{|Z(f_j,R_i,\omega_l)|}} \quad (2)$$

Additionally for further evaluations we calculate the relative deviation of absolute impedance mean value $\delta_{v,|Z(f_j,R_i,\omega_l)|}$ for each sample $v \in [1, 300]$ at each parameter combination:

$$\delta_{v,|Z(f_j,R_i,\omega_l)|} = \frac{|Z_v(f_j,R_i,\omega_l)| - \mu_{|Z(f_j,R_i,\omega_l)|}}{\mu_{|Z(f_j,R_i,\omega_l)|}} \cdot 100\% \quad (3)$$

3 RESULTS

The solid curves in Fig. 3 present the absolute impedance value measured with the SCS at different velocities. Due to the fact, that the curves overlap for each R_i the dependence of the measured impedance mean value from the rotational speed is negligible. The effective measurement range is characterized with the $f_{c,i}$ for each R_i. These bound results from the capacitive coupling in the SCS which can be approximated as a parallel capacity $C_{||}$ with a value of 4pF. Fig. 4 shows the SNR-plains of R_1, R_2 and R_3. If the SCS stands still, the SNR is above 4. The rotation causes an initial decrease of the SNR of at least 1 at R_1 up to at least 3 at R_3, afterwards the SNR decreases slightly with increasing rotational speed. Additionally, the SNR for R_1 decreases for higher frequencies below 2. A full particulars view on this noise behavior gives Fig. 5 with the into vertical pixel lines mapped histograms from the relative deviation of absolute impedance mean value $\delta_{v,|Z(f_j,R_i,\omega_l)|}$. The y-axis of each image shows the accuracy of the measurement depending on the measuring parameters R_i and ω_l. High accuracy regions ($\delta_{v,|Z(f_j,R_i,\omega_l)|} \leq 0.2\%$) can be found in the reference measurements without the SCS, if the SCS doesn't rotates and for lower frequencies at the R_1 test resistor, whereby in the last case the maximum frequency of the high accuracy region decreases. At the higher test resistors we can observe an increasing of the accuracy above the cutoff frequency $f_{c,i}$ (red line), this is caused by the capacitive coupling, which decreases the measured absolute impedance value at higher frequency to impedance regions with higher accuracy.

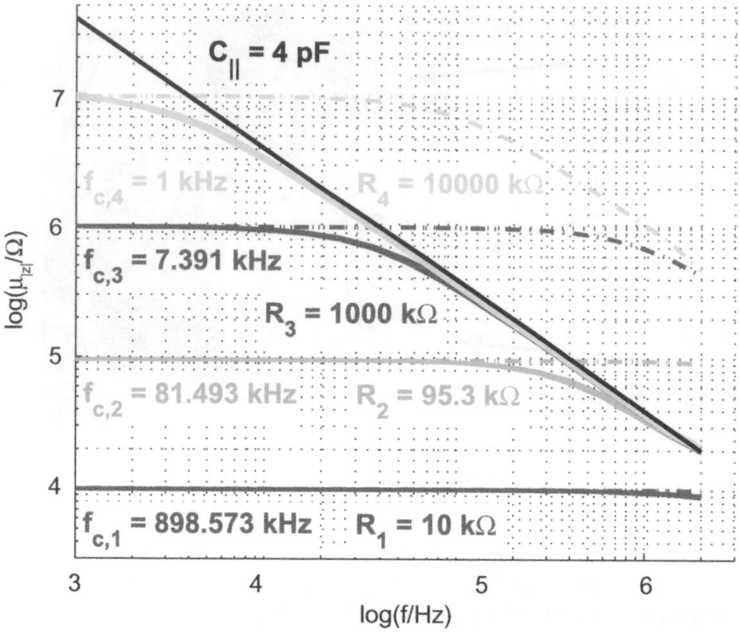
C|| = 4 pF
fc,4 = 1 kHz
R4 = 10000 kΩ
fc,3 = 7.391 kHz
R3 = 1000 kΩ
fc,2 = 81.493 kHz
R2 = 95.3 kΩ
fc,1 = 898.573 kHz
R1 = 10 kΩ
log(μ|z|/Ω)
log(f/Hz)

Figure 3. Frequency depended absolute impedance mean value of test resistors R_i. *Dotted curves*: Ideal resistor curves. *Dash-dot curves*: Reference measurement without the SCS. *Solid curves*: Resistor curves measured with the SCS at different ω. $C_{\|}$ represents the capacitive coupling of the SCS and $f_{c,i}$ is the cutoff frequency of each resistor.

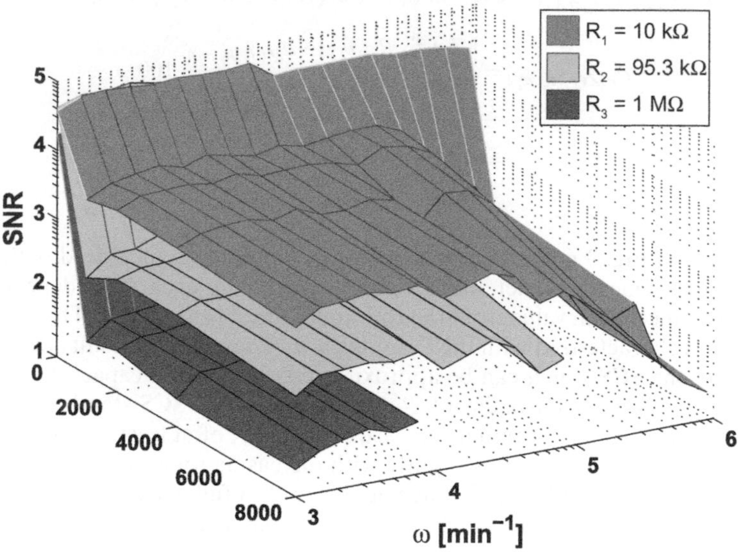
R1 = 10 kΩ
R2 = 95.3 kΩ
R3 = 1 MΩ
SNR
ω [min⁻¹]

Figure 4. SNR-plains of R_1, R_2 and R_3 over the rotational speed ω_l and the frequencies $f_j < f_{c,i}$.

Except for special cases, the Agilent E4890A LCR-Meter measures partial overloads when the SCS rotates, in particular above $f_{c,i}$. But the overloads were also observed frequency independent at $R = 100$ kΩ. These effects are probably originated in the limitations of the used LCR-Meter.

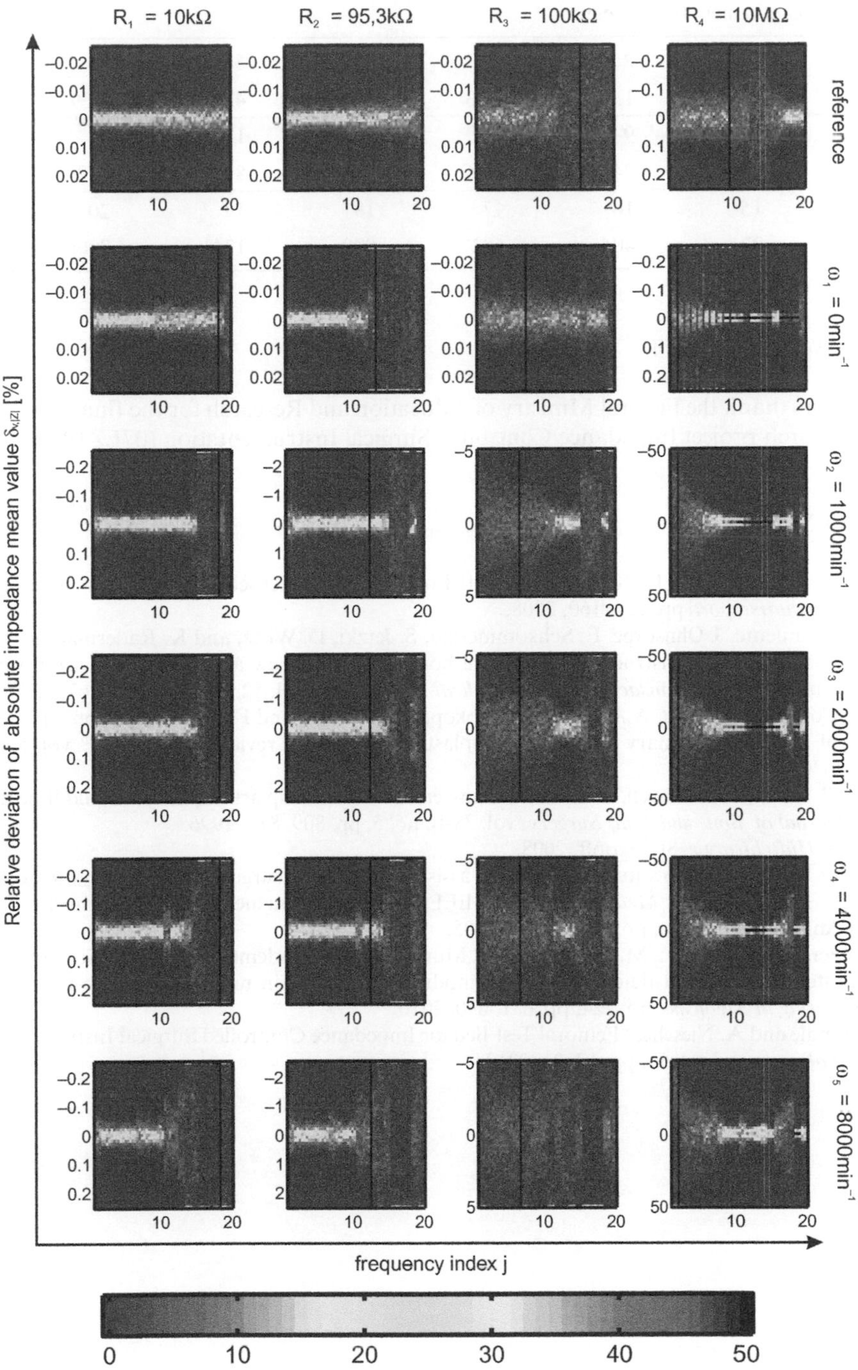
R1 = 10kΩ
R2 = 95,3kΩ
R3 = 100kΩ
R4 = 10MΩ
reference
ω1 = 0min-1
ω2 = 1000min-1
ω3 = 2000min-1
ω4 = 4000min-1
ω5 = 8000min-1
Relative deviation of absolute impedance mean value δ|Z| [%]
frequency index j
0
10
20
30
40
50

Figure 5. Histograms from the relative deviation of absolute impedance mean value over frequencies f_j transliterated into pixel for different test resistors R_i and rotational speed ω_l. The pixels to the right of each vertical cutoff frequency bar represent measurements over the individual cutoff frequency $f_{c,i}$. The histogram bar values are transliterated into pixel corresponding to the bottom color bar, whereby bar values over 50 are combined with the 50 color value. Each vertical pixel line has 50 pixels and represents one histogram from a measurement sweep with 100 values and three histograms are mapped next to each other for each frequency f_j (see Tab. 1).

Table 1. Measurement frequencies.

j	1	2	3	4	5	6	7
f_j [kHz]	1,0	1,49	2,23	3,32	4,95	7,39	11,0
j	8	9	10	11	12	13	14
f_j [kHz]	16,5	24,5	36,6	54,6	81,4	122	181
j	15	16	17	18	19	20	
f_j [kHz]	271	404	602	899	1341	2000	

ACKNOWLEDGMENT

The authors thank the Federal Ministry of Education and Research for the financial support of the research project Impedance Controlled Surgical Instrumentation (07EZ1005).

REFERENCES

[1] O. Boy, S. Hahn, and E. Kociemba, "Hüft-Endoprothesenwechsel und—komponentenwechsel," *BQS Qualitätsreport*, pp. 154–160, 2008.

[2] M. de la Fuente, J. Ohnsorge, E. Schkommodau, S. Jetzki, D. Wirtz, and K. Radermacher, "Fluoroscopy-based 3-d reconstruction of femoral bone cement: A new approach for revision total hip replacement," *IEEE Transactions on Biomedical Engineering*, vol. 52, no. 4, pp. 664–675, 2005.

[3] K.L. Corbett, E. Losina, A.A. Nti, J.J.Z. Prokopetz, J.N. Katz, and F.P. Rannou, "Population-based rates of revision of primary total hip arthroplasty: A systematic review," *PLoS ONE*, vol. 5, no. 10, p. e13520, 2010.

[4] P. Li, P. Ingle, and J. Dowell, "Cement-within-cement revision hip arthroplasty; should it be done?," *The Journal of Bone and Joint Surgery*, vol. 78-B, no. 5, pp. 809–811, 1996.

[5] J. Pfeil, *Hüftchirurgie*. Steinkopff, 2008.

[6] R. Ellis, "From scans to sutures: Computer-assisted orthopedic surgery in the twenty-first century," in *Conf Proc IEEE Eng Med Biol Soc*, no. IEEE Proceedings Engineering in Medicine and Biology 27th Annual Conference, pp. 7234–7237, 2005.

[7] S. Heger, M. Niggemeyer, M. de la Fuente, T. Mumme, and K. Rademacher, "Trackerless ultrasound-integrated bone cement detection using a modular minirobot in revision total hip replacement," *Engineering in Medicine*, no. 224, pp. 681–690, 2010.

[8] C. Brendle and A. Niesche, "Femoral Test Bed for Impedance Controlled Surgical Instrumentation," *Acta Polytechnica*, vol. 52, pp. 17–21, 2012.

Electrochemical systems

Lecture Notes on Impedance Spectroscopy, Volume 4 – Kanoun (ed)
 ISBN 978-1-138-00140-4

Measurement procedure for the dynamic determination of total hardness of water during the washing process

R. Gruden & D. Sanchez
Seuffer GmbH & Co., KG Forschung und Entwicklung, Calw, Germany

O. Kanoun
Technische Universität Chemnitz Professur für Mess- und Sensortechnik, Chemnitz, Germany

ABSTRACT: The total hardness of water which is used for textile washing varies strongly worldwide. It differs partially by a multiple within a municipality. An optimum washing result in a minimum of water and detergent consumption assumes an accurate knowledge of the total hardness of water. This paper discusses a new measurement procedure by which the total hardness of water can be determined during the washing process. Two sensors of different materials were used. The results obtained from the measurement provide the basis for an automated and optimal dosage.

Keywords: Cyclic Voltammetry, Impedance Spectroscopy, total hardness of water

1 INTRODUCTION

Water plays a central role for the washing process. It dissolves and transports detergent and dirt. Furthermore, it transfers thermal and mechanical energy to the laundry [1]. In 2007, each German consumed 15 liters of fresh water on average for textile washing every day [2]. Since fresh water is a rare and precious good, an economical and environmentally-oriented handling is required. Optimal washing with a minimum of water use requires the detection of the important parameters of the washing process. The most important measured variable that water provides for this purpose is the total hardness of water. It is defined as the concentration of the alkaline earth metal ions calcium and magnesium per liter of water $c(Ca^{2+} + Mg^{2+})^{\text{mmol/L}}$ [3]. Alkaline earth metal ions interfere with the washing process because they form with soap or anionic surfactants sparingly soluble salts. These unwanted substances are deposited on the laundry and the heating elements of the washing machine and cause discoloration and damage. In addition, the total hardness of water affects the bleaching process and the formation of foam. To counteract these negative effects, modern detergents contain builders [1, 4, 5]. In Germany, water varies from a total hardness of water of approx. 2° dH (0.356 mmol/L) in Furtwangen (Black Forrest) up to approx. 57° dH (10.146 mmol/L) in Nuremberg-Schweinau [6]. In extreme cases, values between 1° dH (0.178 mmol/L) and 100° dH (17.8 mmol/L) can occur [7]. These numbers show how important the knowledge of the total hardness of water and the resulting detergent dosage is.

The aim of this study is to measure the total hardness of water with a sensor system which is suitable for the measuring principles, Impedance Spectroscopy (EIS) and Cyclic Voltammetry (CV) in combination. In addition, the measurement procedure used is dynamic, affordable and sufficiently precise and does without any environmentally harmful substances.

2 STATE OF THE ART

To determine the total hardness of water, different methods are available but they are not suitable for an inexpensive and dynamic measurement during the washing process. Titrimetric methods such as acid-base titration for determining the *m*- and *p*-values [7] can be carried out only under laboratory conditions and not during the washing process. Such professional and high precision titrimetric methods are very expensive [8]. Voltammetric methods use complex ion-selective electrodes [9] or environmentally harmful mercury electrodes [10–12]. Optical procedures require monochromatic light sources for each species. The optical equipment is sensitive to contamination [13, 14]. Simple and inexpensive structured water hardness measuring instruments, such as conductive measuring devices, determine the total hardness by measuring conductivity [15]. This procedure is inaccurate because the conductivity depends on the total ion concentration and not only on the concentration of alkaline earth metals [7].

This work focuses on laboratory results. The feasibility in washing machines is under test.

3 THEORY

Total hardness of water (TH) is defined as the concentration of alkaline earth metal ions. Since beryllium and radium do not occur and strontium and barium occur only in traces the total hardness is the molar concentration of calcium and magnesium ions in mmol/L [7]. In practice, country-specific definitions are common. For example, in Germany the term "Grad Deutsche Härte" (°dH) is used. 1° dH is equivalent to a concentration of $c(Ca^{2+} + Mg^{2+}) = 0.178$ mmol/L [1, 3]. Water is very individual in its composition depending on its origin. A lot of different dissolved filter materials affects the electrical and chemical properties of water.

In general, the molar conductivity of an aqueous solution depends on the total ion concentration.

$$\Lambda_{\mathrm{mTI}} = \frac{\chi}{c_{\mathrm{TI}}} \tag{1}$$

Here, χ is the conductivity of the water sample and c_{TI} is the total ion concentration. According to Kohlrauschs square root law (research into conductivities of strong and weak electrolytes) the molar limit conductivity Λ^0_{mTI} is defined as

$$\Lambda^0_{\mathrm{mTI}} = \Lambda_{\mathrm{mTI}} + \mathcal{K}\sqrt{c_{\mathrm{TI}}}\,. \tag{2}$$

The coefficient $\mathcal{K}$ depends on the stoichiometry of the electrolyte. The total ion concentration of the water samples used in this study are very low. Thus applies approximately

$$\Lambda^0_{\mathrm{mTI}} \approx \Lambda_{\mathrm{mTI}} \tag{3}$$

hence

$$\Lambda^0_{\mathrm{mTI}} \approx \frac{\chi}{c_{\mathrm{TI}}}. \tag{4}$$

According to Kohlrauschs law of independent migration of ions

$$\Lambda^0_{\mathrm{mTI}} = \upsilon_+ \cdot \lambda_+ + \upsilon_- \cdot \lambda_-, \tag{5}$$

the molar limit conductivity of these water samples has a linear correlation to all contained species [16, 17].

$$\Lambda^0_{mTI} = \upsilon_{Ca^{2+}} \cdot \lambda_{Ca^{2+}} + \upsilon_{Mg^{2+}} \cdot \lambda_{Mg^{2+}} + \upsilon_{Na^+} \cdot \lambda_{Na^+} + \upsilon_{HCO_3^-} \cdot \lambda_{HCO_3^-} + \upsilon_{Cl^-} \cdot \lambda_{Cl^-} + \upsilon_{SO_4^{2-}} \cdot \lambda_{SO_4^{2-}} \quad (6)$$

υ_+ and υ_- are the number of cations and anions (e.g. $\upsilon_+ = 2$ for Ca^{2+} and $\upsilon_- = 1$ for HCO_3^-) per unit formula, λ_- and λ_- are the molar limit conductivity of the cations or anions per unit formula. Relevant to the total hardness of water are the cations Ca^{2+} and Mg^{2+} and the corresponding anions. These can be all types of negative species which occur in water, mainly HCO_3^-, Cl^- and SO_4^{2-}. The molar limit conductivity of total hardness Λ^0_{mTH} is a share of the molar limit conductivity of the total ion concentration Λ^0_{mTI}. The relationship is linear.

$$\Lambda^0_{mTH} = \alpha \cdot \Lambda^0_{mTI} \quad (0 < \alpha < 1) \quad (7)$$

α is the percentage of the total ion concentration which is responsible for the total hardness of water. α depends on the concentration of Ca^{2+} and Mg^{2+} and the concentration of the corresponding anions. Only the calcium and magnesium ions and a part of the anions contribute to the conductivity of total hardness of water.

$$\begin{aligned} \Lambda^0_{mTH} = {} & \upsilon_{Ca^{2+}} \cdot \lambda_{Ca^{2+}} + \upsilon_{Mg^{2+}} \cdot \lambda_{Mg^{2+}} + \alpha_{HCO_3^-} \cdot \upsilon_{HCO_3^-} \cdot \lambda_{HCO_3^-} \\ & + \alpha_{Cl^-} \cdot \upsilon_{Cl^-} \cdot \lambda_{Cl^-} + \alpha_{SO_4^-} \cdot \upsilon_{SO_4^{2-}} \cdot \lambda_{SO_4^{2-}} \end{aligned} \quad (8)$$

α_{anion} is the percentage of anions corresponding to the cations Ca^{2+} and Mg^{2+}.

4 EXPERIMENTAL

The test media were five synthetic water samples made according to EN 60734:2003 Method B [3] with different degrees of hardness. All samples are qualitatively identical (see table 1). The required high purity water was prepared with a Siemens LaboStar UV7 immediately before the experimental procedure. The reference measurement of each water was carried out with the titrator TA20plus and the software TitriSoft 2.6 of SI Analytics (results see table 1). For the EIS- and CV-measurements, a Zahner Zennium electrochemical workstation was used. During the measurements, the temperature of the test medium was kept constant using a Julabo LH46 temperature control in a silicone oil-cooled reactor vessel at (24 ± 0.02) °C. Two home-made sensors of different material combinations were used. Each test water was investigated with both sensors and both measuring principles.

Table 1. List of the synthetic water samples (measured values).

Water sample	Total hardness (mmol/L)	Total ion concentration (mmol/L)	$c(Ca^{2+})$ (mg/L)	$C(Mg^{2+})$ (mg/L)	χ (μS/cm)
W1	0.5	2.71	15.13	3.02	198
W2	1.0	5.42	30.37	5.98	396
W3	1.5	8.12	44.89	9.21	594
W4	2.0	10.83	59.81	12.38	792
W5	2.51	13.54	73.81	16.24	990

This resulted in 20 sub-tests. The following was done with each water sample.

1. Preparing the water sample.
2. Reference measurement by titration.
3. EIS measurement with sensor 1.
4. CV measurement with sensor 1.
5. EIS measurement with sensor 2.
6. CV measurement with sensor 2.

5 RESULTS AND DISCUSSION

5.1 *Impedance Spectroscopy (EIS)*

Fig. 1 and Fig. 2 show the impedance spectra of the two sensors. Each spectrum represents a water sample.

The local maximum turning points in Fig. 1 correspond to the conductance G of the measuring cells. The conductance depends on the sensor geometry, expressed in the cell constant K, and the conductivity χ of the corresponding electrolyte (in this case the test water samples).

$$G = \chi \cdot K \tag{9}$$

From the calculated conductivities χ of each test water (see table 1) and the measured conductance values G resulted in the following arithmetic averages and standard deviation for the cell constants.

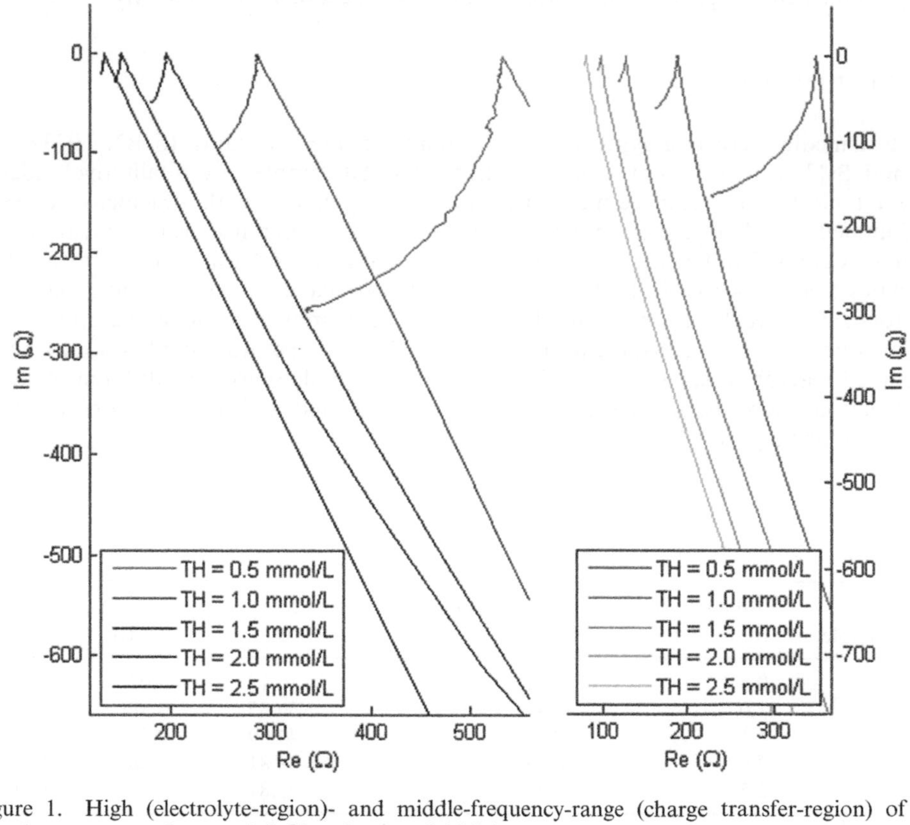
0
-100
-200
-300
-400
-500
-600
Im (Ω)
TH = 0.5 mmol/L
TH = 1.0 mmol/L
TH = 1.5 mmol/L
TH = 2.0 mmol/L
TH = 2.5 mmol/L
200
300
400
500
Re (Ω)
0
-100
-200
-300
-400
-500
-600
-700
Im (Ω)
TH = 0.5 mmol/L
TH = 1.0 mmol/L
TH = 1.5 mmol/L
TH = 2.0 mmol/L
TH = 2.5 mmol/L
100
200
300
Re (Ω)

Figure 1. High (electrolyte-region)- and middle-frequency-range (charge transfer-region) of the impedance spectra of sensor 1 (left) and sensor 2 (right).

Sensor 1: $K_1 = 0.117$ cm; $s_1 = 0.0086$ cm
Sensor 2: $K_2 = 0.075$ cm; $s_2 = 0.0034$ cm

Slight variations in cell constants can be observed for both sensors. The cell constant of Sensor 2 has the stabler factor.

The total hardness of all test water samples behaved linearly to the total ion concentration because all water samples were qualitatively identical. The molar limit conductivity of every water sample was $\Lambda^0_{mTI} = 7.31 \mathrm{mSm}^2 \mathrm{mol}^{-1}$ (see table 1). The expected linear correlation between total hardness of water and conductivity could be confirmed by the results of impedance spectroscopy (see Fig. 3).

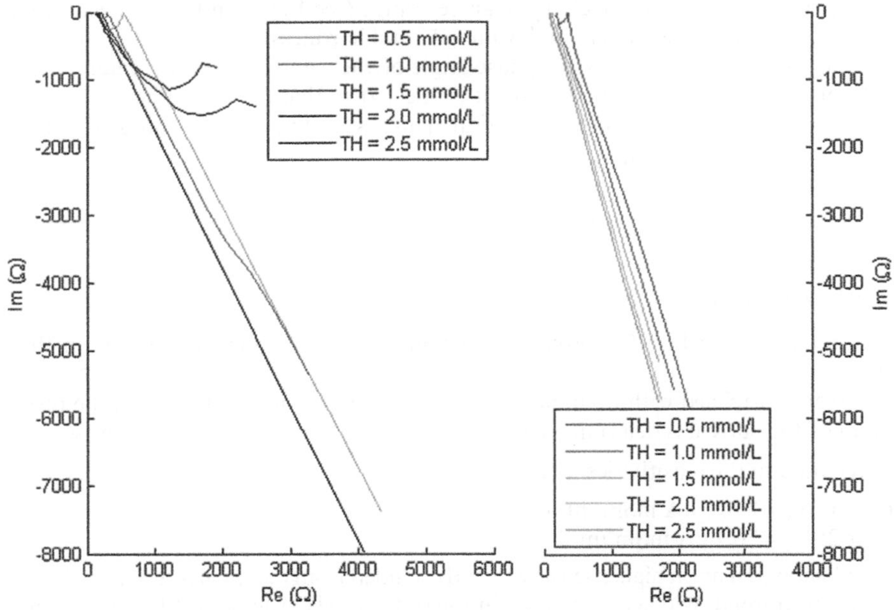

TH = 0.5 mmol/L
TH = 1.0 mmol/L
TH = 1.5 mmol/L
TH = 2.0 mmol/L
TH = 2.5 mmol/L
Im (Ω)
Re (Ω)

Figure 2. Middle (charge transfer-region)- and low-frequency-range (mass transport-region) of the impedance spectra of sensor 1 (left) and sensor 2 (right).

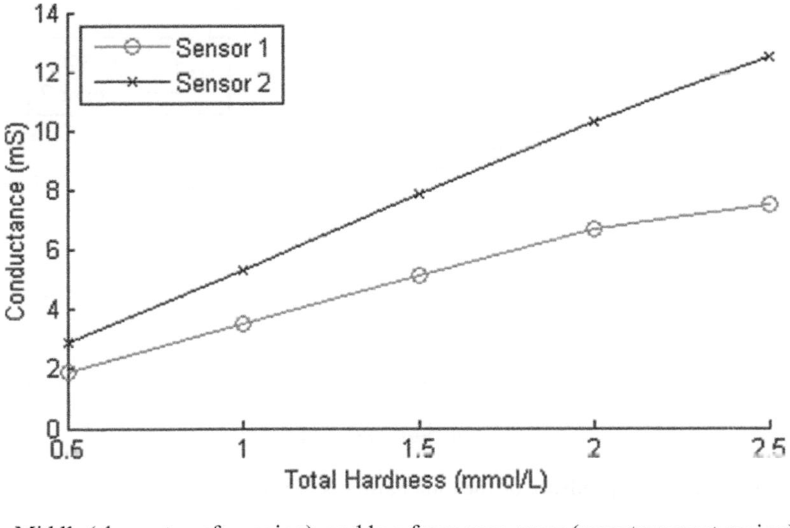

Sensor 1
Sensor 2
Conductance (mS)
Total Hardness (mmol/L)

Figure 3. Middle (charge transfer-region)- and low-frequency-range (mass transport-region) of the of the impedance spectra of sensor1 (left) and sensor 2 (right).

The measurements resulted in the following sensitivities S_c[1].

Sensor 1: $S_{c1} = 2.89$ mSm³ mol⁻¹

Sensor 2: $S_{c2} = 4.83$ mSm³ mol⁻¹

Sensor 2 showed a higher sensitivity and a more linear characteristic curve. The total hardness of water affects the characteristic ranges charge transfer and diffusion (see "Three Regions Model" [18–20]). Fig. 2 shows the large influence of the total hardness of the diffusion region of the characteristic curve of Sensor 1. The characteristic curves of the total hardness 1.5 and 2.0 mmol/L were particularly pronounced. These characteristics have not been studied in detail. The characteristic curves of Sensor 2 in the diffusion range are qualitatively identical. They quantitatively correlate with the total hardness of water unsystematically. The influence on the charge transfer range (see Fig. 1 and Fig. 2) was greater at low value of total hardness than at high value of total hardness.

Fig. 4 shows the dependence of the phase angle of the total hardness of water.

The frequency for the phase measurement is selected so that the measurement results were linear and reproducible at maximum resolution. This resulted in the following sensitivities S_p.

Sensor 1: $S_{p1} = -10.96$ °m³ mol⁻¹

Sensor 2: $S_{p2} = -14.53$ °m³ mol⁻¹

Sensor 2 showed a higher sensitivity and a more linear characteristic curve.

5.2 *Cyclic Voltammetry (CV)*

The voltammogram in Fig. 5 showed the current response of the two sensors in the time domain.

The current correlated with total hardness of water. To compare the results, the respective maximum of the current is determined and applied to the total hardness of water (see Fig. 6).

This resulted in the sensitivities S_i.

Sensor 1: $S_{i1} = 896.6$ μA mSm³/mol

Sensor 2: $S_{i2} = 26.1$ μA mSm³/mol

Sensor 1 showed a much higher sensitivity and excluded from the smallest total hardness value a perfectly linear trend. With a required resolution of the total hardness of 0.1° dH (17.8 μmol/L) the minimum resolution of the different sensor parameters (listed in table 2) are necessary.

With a suitable choice of the sensors to the corresponding measurements, the circuit complexity and cost can be minimized.

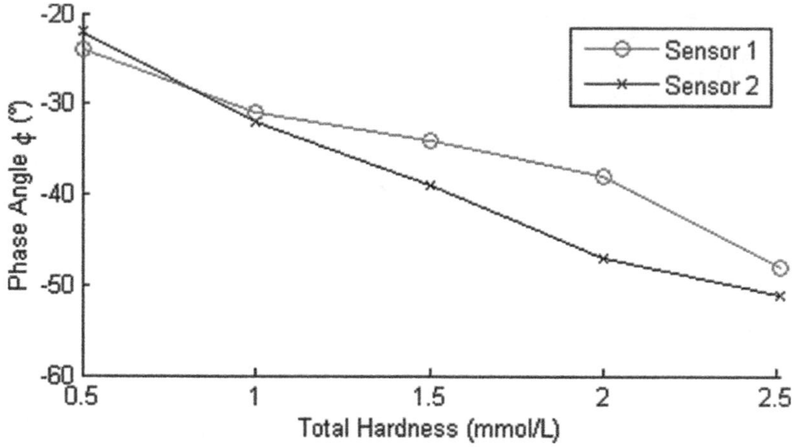

Sensor 1
Sensor 2
Phase Angle ϕ (°)
-20
-30
-40
-50
-60
0.5
1
1.5
2
2.5
Total Hardness (mmol/L)

Figure 4. Phase angle in dependence of the total hardness of water.

[1]All sensitivities in this work were determined by a linear regression calculation.

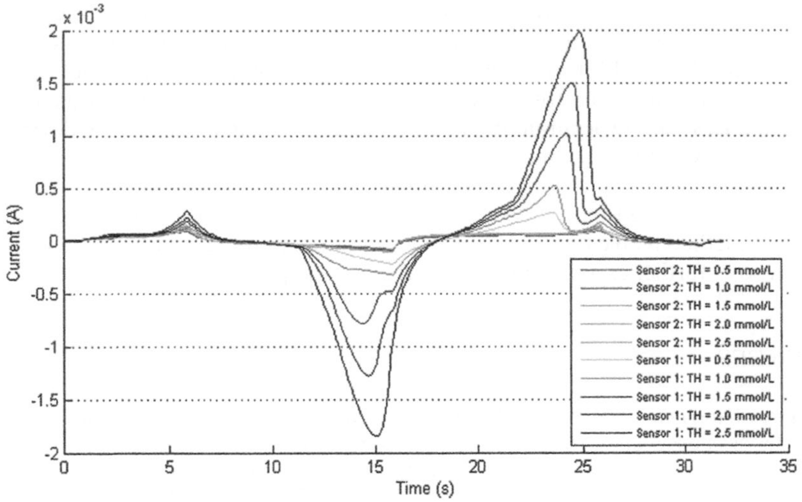

x 10^-3
Current (A)
Time (s)
Sensor 2: TH = 0.5 mmol/L
Sensor 2: TH = 1.0 mmol/L
Sensor 2: TH = 1.5 mmol/L
Sensor 2: TH = 2.0 mmol/L
Sensor 2: TH = 2.5 mmol/L
Sensor 1: TH = 0.5 mmol/L
Sensor 1: TH = 1.0 mmol/L
Sensor 1: TH = 1.5 mmol/L
Sensor 1: TH = 2.0 mmol/L
Sensor 1: TH = 2.5 mmol/L

Figure 5. Current response of both sensors according to total hardness of water.

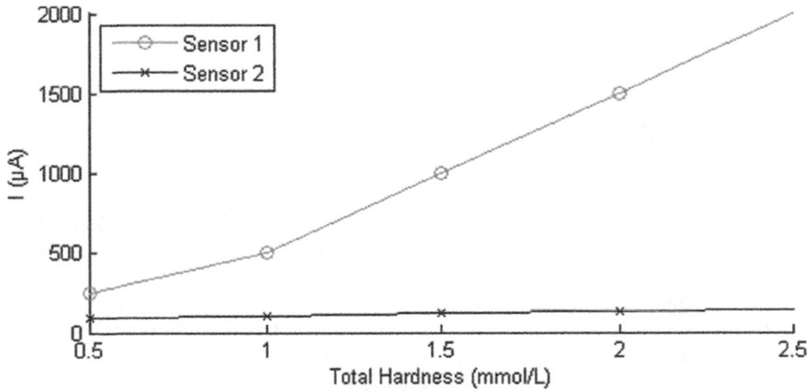

Sensor 1
Sensor 2
I (µA)
Total Hardness (mmol/L)

Figure 6. Current in dependence of the total hardness of water.

Table 2. Minimum resolutions of the sensor parameters by a required result resolution of 0.1 ° dH (17.8 µmol/L).

Sensor	Conductance (µS)	Phaseangle (°)	Current (µA)
S1	50	0.21	155
S2	85	0.26	4.7

6 CONCLUSION

Conductance, phase angle and current response correlated almost linearly with the total hardness of water. These measurement values reacted differently to a change in the total hardness of water which provides a more detailed analysis in view. Sensitivity and linearity of both sensors are strongly dependent on the applied frequency range. Sensor 1 is more suitable for use in low and Sensor 2 for use at higher frequency ranges. Both measuring principles (EIS and CV) disclosed results that complemented each other. The frequency range (in the cyclic voltammetry frequencies in the mHz range were used) and the different forms of presentation (the image of EIS is displayed in the frequency domain, the image of CV is displayed in the time domain) are important in this context. With a reasonable combination of both measurement principles and both sensors, measurement methods to investigate further water parameters can be developed. With this additional information the total hardness of water can be determined more accurately independent of the total ion concentration. The conductivity measurement with the help of the EIS is more accurate than conventional conductive measuring devices because the exact frequency of the local high point and thus, the exact value of the conductivity can be determined. EIS is assumed to provide additional information to separate the influence of the ions which have no influence on the water hardness. Therefore it can be assumed that this new measurement method is able to disclose further important parameters for the washing process. For this purpose, further experiments are in progress and planned.

REFERENCES

[1] G. Wagner, *Waschmittel*. Weinheim: WILEY-VCH, 4., vollständig überarbeit-ete auflage ed., 2010.

[2] "Statistisches bundesamt, www.destatis.de," aufgerufen 10/2009.

[3] "Household electrical applicances performance—hard water for testing (iec 60734:2003)."

[4] G. Jakobi and A. Löhr, *Detergents and Textile Washing*. Weinheim: VCH, 1987.

[5] E. Smulders, W. Rähse, W. von Rybinski, J. Steber, E. Sung, and F. Wiebel, *Laundry Detergents*. Wil, 2002.

[6] "www.wasserhärte.net/stats," aufgerufen 08/2012.

[7] L.A. Hütter, *Wasser und Wasseruntersuchung*. Frankfurt a. M., Aarau: Otto Salle Verlag and Verlag Sauerländer, 1990.

[8] *Titrator TitroLine alpha plus*, Dezember 2003.

[9] K. Cammann and H. Galster, *Das Arbeiten mit ionenselektiven Elektroden*. Berlin, Heidelberg: Springer, dritte auflage ed., 1996.

[10] E. Pungor, G. Nagy, and Z. Feher, "The flat surfaced membrane coated mercury electrode as analytical tool in the continuous voltammetric analysis," *Journal of Electroanalytical Chemistry*, vol. 75, pp. 241–254, 1977.

[11] H.-X. Zhao, W. Cai, D. Ha, H. Wan, and P. Wang, "The study on novel microelectrode array chips for the detection of heavy metals in water pollution," *Journal of Innovative Optical Health Sciences*, vol. 5, pp. 1150002–1–1150002–7, 2012.

[12] A.W. Bott, "Stripping voltammetry," *Current Separations*, vol. 12, pp. 141–147, 1993.

[13] F. Tiersch, "Messtechnik, wasseranalyse," Vorlesungsskript, Hochschule München, k.J.

[14] H. Lang, "Analytisches praktikum," Vorlesungsskript, TU Chemnitz, k.J.

[15] "Syr haustechnik, http://www.syr.de," aufgerufen 15.08.2012.

[16] C.H. Hamann and W. Vielstich, *Elektrochemie*. Weinheim: Wiley-VCH, 3., vollständig überarbeitete und aktualisierte auflage ed., 2005.

[17] P.W. Atkins and J. de Paula, *Physikalische Chemie*. Weinheim: WILEY-VCH, Vierte, vollständig überarbeitete Auflage ed., 2006.

[18] R. Gruden, O. Kanoun, and U. Tröltzsch, "Influence of surface effects on the characteristic curves of detergent sensors," in *9th International Multi-Conference on Signals, Sensors and Devices, 20.-23. März 2012, Chemnitz*, 2012.

[19] P. Kurzweil and H.-J. Fischle, "A new monitoring method for electrochemical aggregates by impedance spectroscopy," *Journal of Power Sources*, vol. 127, pp. 331–340, 2004.

[20] E. Barsoukov and J.R. Macdonald, *Impedance Spectroscopy*. Hoboken: WILEY-INTERSCIENCE, second edition ed., 2005.

Lecture Notes on Impedance Spectroscopy, Volume 4 – Kanoun (ed)
© 2014 Taylor & Francis Group, London, ISBN 978-1-138-00140-4

Impedance measurements at sol-gel-based polysiloxane coatings on aluminum and its alloys

A.A. Younis
National Institute of Standards (NIS), Egypt

W. Ensinger
Department of Materials Science, Technische Universität Darmstadt, Darmstadt, Germany

R. Holze
Institut für Chemie, Technische Universität Chemnitz, Chemnitz, Germany

ABSTRACT: The corrosion protection capabilities of sol-gel-based polysiloxane coatings applied to surfaces of pure aluminum and of aluminum alloy AA7075 prepared from different starting molecules have been evaluated with electrochemical methods. A rating of the studied polymer coatings could be deduced and correlated with the hydro-phobicity of these coatings.

Keywords: aluminum, aluminum alloy, sol-gel, polysiloxane, corrosion

1 INTRODUCTION

The large heat of formation $\Delta H_0 = -1675$ kJ/mol of aluminum oxide is the driving force of the formation of a surface layer on every piece of aluminum exposed to air (or precisely: to oxygen). This layer provides some protection against corrosion. Unfortunately it is not sufficiently stable in the presence of pitting agents like e.g. halide ions [1–4]; the mechanism of the subsequent corrosion has been studied [5–10]. In light-weight construction (aircraft, cars etc.) pure aluminum is rarely used because of insufficient mechanical properties, instead numerous alloys containing a variety of further chemical elements are frequently used in a rapidly growing extent. This growth is related to weight reductions and the associated fuel savings. Unfortunately these alloys are significantly less stable towards corrosion, this is due—among several factors—to the formation of corrosion elements at locations of differing metal alloy composition.

Consequently corrosion protection of aluminum alloy surfaces is needed. Materials employed are those also used for other metals. Inorganic coatings containing chromate (as in zinc chromate [8, 11–13], molybdate as zinc-phosphate-molybdate composite and sodium molybdate ($Na_2MoO_4 \cdot 2H_2O$) [14–17] or tungstate [11, 13], organic compounds having polar groups, such as oxygen, sulfur, and nitrogen [18–26], heterocyclic compounds containing functional groups and/or conjugated double bonds [26–29] and self-assembled layers [30] have been studied as corrosion inhibitors. Because coatings containing heavy metals are already phased out or will be legally limited in their application further options are explored. They include coatings formed by sol-gel processes employing siloxanes as starting compounds [31]. Because polysiloxane coatings do not show self-healing properties *per se* embedding inhibitors in the coatings has been proposed [31–33].

Here we report results of a comparative study of corrosion of aluminum and an aluminum alloy (AA7075) using three different siloxanes with the aim to determine their efficiency and the reasons for differences in coating properties.

2 EXPERIMENTAL

2.1 *Materials*

TEOS (Alfa Aesar, 99%), PTES (Alfa Aesar, 98%), PTMS (Alfa Aesar, 97%) were used as received: Acetic acid (Alfa Aesar, 99%) was employed as a catalyst, acetyl acetone (Alfa Aesar, 99%) helped in moderating the reaction speed, and *n*-propanol (C_3H_7OH, Fisher, 99.9%) was used as a cosolvent to prevent precipitation with water. Cylindrical discs (3 cm diameter and 0.3 cm thickness) cut from rods of pure aluminium (98%) and aluminium alloy (AA7075) were used as electrodes. They were treated on a Beta grinder/polishing machine with SiC abrasive paper 600; cleaned and degreased ultrasonically with isopropanol, washed again with ethanol and dried with air prior to the spin-coating process. Coating solutions were prepared by mixing 2.5 ml TEOS, PTES or PTMS, 1.25 ml acetic acid as a catalyst and 1.25 ml acetyl acetone as stabilizing agent to prevent rapid condensation under constant stirring. Finally a mixture of 2 ml n-propanol and 0.5 ml distilled water was added. The solution was stirred at ambient temperature for 90 min (for a study of the influence of hydrolysis time see [34], the time employed here has been identified as the optimum time—as with the heat treatment temperature discussed below) and deposited onto the samples from a syringe for spin coating. The spinning speed was set to 4000 rpm for 90s. Finally the specimens were annealed in a furnace for 150 min at T = 573 K. Based on a comparative study of corrosion data using the experimental approach described here in detail this temperature has been identified as yielding the best protective properties in terms of E_{corr} and j_{corr}, for further details of the influence of heat treatment duration and temperature see [35]. Thickness of the coatings was determined at an edge to an un-coated part of the specimen to be 89 nm using a surface profilometer (DEKTAK 8000).

A three-electrode electrochemical cell with samples mounted in a holder exposing a surface area of 2 cm^2 (all data are converted to specific data related to 1 cm^2), an Ag/AgCl reference electrode (in saturated KCl-solution) and platinum counter electrode were used. Potentiodynamic current potential curves were recorded by changing the electrode potential from $E_{Ag/AgCl} = -1.5$ V to 0.0 V at a scan rate of 50 mV/s. This scan rate was selected as a compromise between a very low scan rate yielding data very close to equilibrium, almost stationary conditions fraught with significant corrosion damage of the sample, and higher rates with less damage but also with less representative results. The electrolyte solution was 0.05 mol/1 NaCl at pH = 7 at ambient temperature. Potentiodynamic scans were only applied to obtain E_{corr} and to estimate the intrinsic porosity and the protection performance of the films from j_{diss} using a Princeton Applied Research potentiostat (Parstat 2273) with software Power Suite 2.58.

Porosities of the coatings were calculated from dissolution current densities j_{diss} at $E = E_{OCP}$ (open circuit potential) +50 mV. The average film porosity can be calculated from the ratio of the dissolution current densities of coated and uncoated reference samples [36].

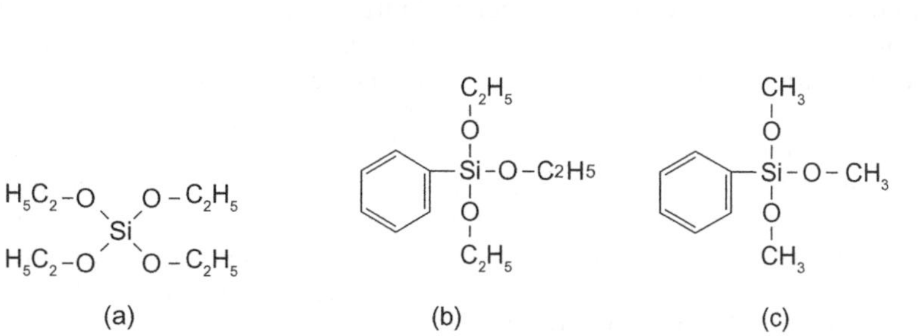

(a)
(b)
(c)

Figure 1. (a) Tetraethoxysilane $Si(OC_2H_5)_4$ (TEOS), (b) phenyltriethoxysilane $PhSi(OC_2H_5)_3$ (PTES) and (c) phenyltrimethoxysilane $PhSi(OCH_3)_3$ (PTMS).

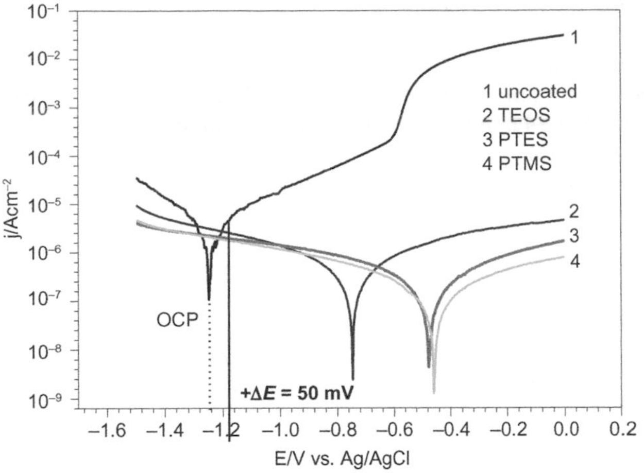
1 uncoated
2 TEOS
3 PTES
4 PTMS
OCP
+ΔE = 50 mV
j/Acm⁻²
E/V vs. Ag/AgCl

Figure 2. Potentiodynamic polarization curves recorded in aqueous solution of 0.05 mol/l NaCl of pure aluminium uncoated and spin-coated with TEOS, PTES and PTMS, additional line indicates exemplarily the electrode potential were j_{diss} was obtained. OCP: open circuit potential (= E_{corr}).

This method has been successfully applied on different coating systems before and is described in detail elsewhere [37].

Values of the resistance R_p representative of the rate of the corrosion reaction were obtained with Linear Polarization Resistance (LPR) and Electrochemical Impedance Measurements (EIS). These resistances are related to the polarization resistance by the Stern-Geary linear approximation of the Butler-Volmer-equation

$$R_p = \frac{B}{I_{corr}} \tag{1}$$

with the corrosion current I_{corr} and the Stern-Geary constant B calculated according to

$$B = \frac{B_a B_c}{2.303(B_c - B_a)} \tag{2}$$

using the Tafel slopes B_a and B_c of the anodic and cathodic parts of Tafel plots in mV/dec. LPR measurements were performed at $dE/dt = 2.0$ mV/s with the samples previously immersed for 24 h in 0.05 mol/l NaCl solution. Obtained Tafel plots did not show sufficiently wide linear regions, thus for the sake of comparing the different coatings and metals within this report slopes of 100 mV/dec were estimated. Corrosion inhibition effiencies were calculated according to

$$I_E = \frac{j_{corr,uncoated} - j_{corr,coated}}{j_{corr,uncoated}} \cdot 100 \quad \text{and} \quad I_E = \frac{R_{p,coated} - R_{p,uncoated}}{R_{p,coated}} \cdot 100 \tag{3}$$

with the polarization resistance $R_{p,coated}$ of the coated electrode and the uncoated electrode $R_{p,uncoated}$ and the respective meanings of j_{corr} [38].

Impedance data were obtained using a sine wave stimulus of 10 mV amplitude between 100 kHz and 39.8 mHz with 10 points per frequency decade and the instrumental setup described above.

3 RESULTS AND DISCUSSION

Polarization curves of uncoated and coated specimens of aluminium and its alloy in aqueous electrolyte solution of 0.05mol/l NaCl are shown in Fig. 3 and 4, respectively.

Results obtained after 24 h immersion of the alloy sample in an aqueous solution 0.05 mol/l NaCl are shown below in Fig. 4.

In case of the uncoated sample exposure to the electrolyte solution has apparently resulted in formation of a slightly more protective layer than formed immediately after contact with

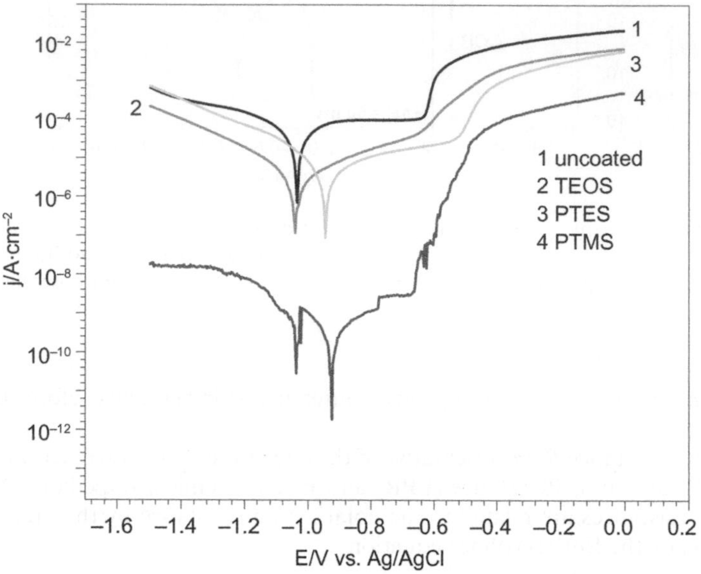

1
2
3
4
1 uncoated
2 TEOS
3 PTES
4 PTMS
j/A·cm⁻²
10⁻²
10⁻⁴
10⁻⁶
10⁻⁸
10⁻¹⁰
10⁻¹²
−1.6 −1.4 −1.2 −1.0 −0.8 −0.6 −0.4 −0.2 0.0 0.2
E/V vs. Ag/AgCl

Figure 3. Potentiodynamic polarization curves recorded in aqueous solution of 0.05 mol/l NaCl of Al alloy uncoated and spin-coated with TEOS, PTES and PTMS.

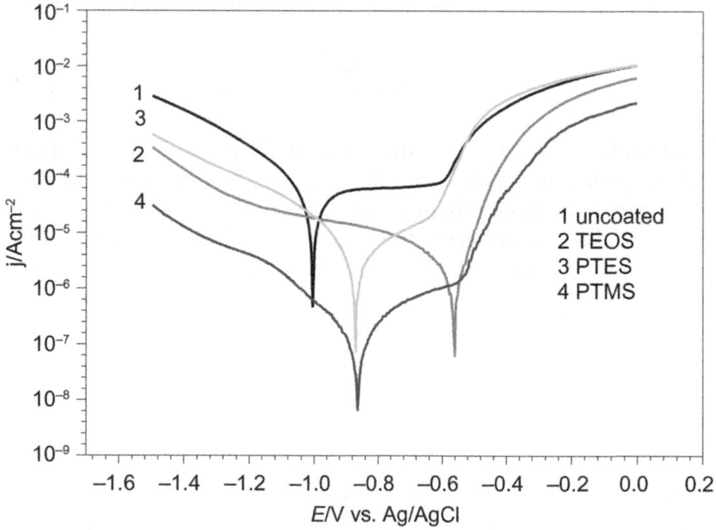

1
3
2
4
1 uncoated
2 TEOS
3 PTES
4 PTMS
j/Acm⁻²
10⁻¹
10⁻²
10⁻³
10⁻⁴
10⁻⁵
10⁻⁶
10⁻⁷
10⁻⁸
10⁻⁹
−1.6 −1.4 −1.2 −1.0 −0.8 −0.6 −0.4 −0.2 0.0 0.2
E/V vs. Ag/AgCl

Fgiure 4. Potentiodynamic polarization curves recorded in aqueous solution of 0.05 mol/l NaCl for aluminium alloy uncoated and coated with TEOS, PTES and PTMS after immersion for 24 h in aqueous solution 0.05 mol/l NaCl.

the sodium chloride solution (compare Table 1) evident in a slightly more positive value of E_{corr} and a smaller value of j_{diss}, in case of the coated samples values of j_{diss} have increased slightly indicating some degradation of the coatings. The values of E_{corr} all moved into positive direction confirming the generally protective properties of the coating.

Corrosion current densities and values of R_p can be obtained from LPR and impedance measurements.

In a detailed study of the influence of pH-value and chloride ion concentration [39] the effect of the former parameter on stability of the alloy was verified.

Again the protective effect of the coating is evident, in particular at extreme pH-values. PTMS was selected for the latter experiments because its superior protection capability had become evident. As a reason for this superiority materials properties of the polysiloxane coatings have to be considered. Because the coatings act solely as a mechanical barrier against contact of the metal (or metal oxide) surface with any corrosive medium wettability (hydrophobicity) may play a major role. The more hydrophobic a coating the better its protection because disabling contact becomes more effective with growing hydrophobicity.

Table 1. Corrosion potentials (E_{corr}), dissolution current densities (j_{diss}) and porosities obtained from potentiodynamic polarization curves shown in Fig. 3 and 4 for uncoated and coated samples immediately after immersionin aqueous solution of 0.05 mol/l NaCl are collected below.

	E_{corr} in mV vs. Ag/AgCl		j_{diss} µA/cm^2		Porosity %	
Specimens	Pure Al	Al alloy	Pure Al	Al alloy	Pure Al	Al alloy
Blank	–1250	–1043	3,47	52	100	100
TEOS	–765	–1043	0,33	4	9,5	7,7
PTES	–477	–950	0,16	5,61	4,6	10
PTMS	–462	–912	0,10	0,0006	2,9	0,1

Table 2. Dissolution current densities (j_{diss}) and corrosion potentials (E_{corr}) for aluminium alloy uncoated and coated with TEOS, PTES and PTMS, after immersion for 24 h in aqueous solution 0.05 mol/l NaCl.

Coating	E_{corr} in mV vs. Ag/AgCl	j_{diss} µA/cm^2
Blank	–1009	40.7
TEOS	–560	8.7
PTES	–870	4.3
PTMS	–860	0.2

Table 3. Electrochemical results of linear polarization measurements and corrosion inhibition efficiencies for Al alloy in aqueous solution of 0.05 mol/l NaCl.

	E_{corr} V	R_p Ω/cm^2	j_{corr} µA/cm^2	Corrosion rate in mm/yr	I_E %
Uncoated	–0.582	1.6×10^3	2.17×10^{-5}	9.0×10^{-5}	–
TEOS	–0.689	10×10^3	2.17×10^{-6}	5.4×10^{-7}	89
PTES	–0.678	13×10^3	1.6×10^{-6}	4.4×10^{-7}	92
PTMS	–0.677	159×10^3	1.36×10^{-7}	1.9×10^{-7}	99

Table 4. LPR and EIS (marked *) results for uncoated and coated specimens after immersion in electrolyte solutions with different pH-values after immersion for 10 min (LPR) or 1 d (EIS).

pH	E_{corr} mV	j_{corr} µA/cm²	$R_p/10^3$ Ω/cm²	Porosity %
	Uncoated			
0.85	−621	2010	0.06	–
3	−336	50.7	1.38	–
7	−1010	34.4	1.68	–
7	–	14*	1*	–
10	−1040	36.5	1.36	–
12.85	−1340	1560	0.01	–
	Coated with PTMS			
0.85	−369	4.0	1	0.2
3	−261	4.6	27	9.1
7	−870	0.18	5214	0.5
7	–	–	159*	0.14*
10	−663	1.87	600	5.1
12.85	−1212	46.3	2	4.5

Contact angle measurements for the respective layer reported elsewhere [40, 41] suggest a hydrophobicity rating with the contact angle of PTMS (called elsewhere PTMOS) of 96 ± 1° being substantially larger than 88 ± 1° for TEOS and 90 ± 1° for PTES (called PhTEOS elsewhere). This rating matches the efficiency. Efficiency at all pH-values increases from TEOS-coating to a maximum with PTMS-coating. The lower efficiency of the TEOS-based coating may be due to an increased solubility of the protective layer caused by formation of hydrogen bonds between protons from water molecules and OH-groups on the Al surface facilitated by the lower hydrophobicity of this layer as compared to that of a PTES or—even more—PTMS-layer.

REFERENCES

[1] C. Brett, "On the electrochemical behaviour of aluminium in acidic chloride solution," *Corrosion science*, vol. 33, no. 2, pp. 203–210, 1992.

[2] R. Ambat and E. Dwarakadasa, "Studies on the influence of chloride ion and pH on the electrochemical behaviour of aluminium alloys 8090 and 2014," *Journal of Applied Electrochemistry*, vol. 24, no. 9, pp. 911–916, 1994.

[3] C. Brett, "The application of electrochemical impedance techniques to aluminium corrosion in acidic chloride solution," *Journal of Applied Electrochemistry*, vol. 20, no. 6, pp. 1000–1003, 1990.

[4] P. Cabot, F. Centellas, J. Garrido, E. Perez, and H. Vidal, "Electrochemical study of aluminium corrosion in acid chloride solutions," *Electrochimica acta*, vol. 36, no. 1, pp. 179–187, 1991.

[5] A. Mazhar, W. Badawy, and M. Abou-Romia, "Impedance studies of corrosion resistance of aluminium in chloride media," *Surface and Coatings Technology*, vol. 29, no. 4, pp. 335–345, 1986.

[6] R. Foley and T. Nguyen, "The Chemical Nature of Aluminum Corrosion V. Energy Transfer in Aluminum Dissolution," *Journal of The Electrochemical Society*, vol. 129, no. 3, pp. 464–467, 1982.

[7] L. Tomcsanyi, K. Varga, I. Bartik, H. Horányi, and E. Maleczki, "Electrochemical study of the pitting corrosion of aluminium and its alloys – II. Study of the interaction of chloride ions with a passive film on aluminium and initiation of pitting corrosion," *Electrochimica Acta*, vol. 34, no. 6, pp. 855–859, 1989.

[8] W. Badawy, F. Al-Kharafi, and A. El-Azab, "Electrochemical behaviour and corrosion inhibition of Al, Al-6061 and Al–Cu in neutral aqueous solutions," *Corrosion Science*, vol. 41, no. 4, pp. 709–727, 1999.
[9] F. Hunkeler, G. Frankel, and H. Bohni, "Technical Note: On the Mechanism of Localized Corrosion," *Corrosion*, vol. 43, no. 3, pp. 189–191, 1987.
[10] N. Sato, "The stability of localized corrosion," *Corrosion Science*, vol. 37, no. 12, pp. 1947–1967, 1995.
[11] S.S.A. Rehim, H.H. Hassan, and M.A. Amin, "Corrosion and corrosion inhibition of Al and some alloys in sulphate solutions containing halide ions investigated by an impedance technique," *Applied Surface Science*, vol. 187, no. 3, pp. 279–290, 2002.
[12] C. Brett, I. Gomes, and J. Martins, "The electrochemical behaviour and corrosion of aluminium in chloride media. the effect of inhibitor anions," *Corrosion Science*, vol. 36, no. 6, pp. 915–923, 1994.
[13] S. El Abedin, "Role of chromate, molybdate and tungstate anions on the inhibition of aluminiumin chloride solutions," *Journal of Applied Electrochemistry*, vol. 31, no. 6, pp. 711–718, 2001.
[14] P. Natishan, E. McCafferty, and G. Hubler, "Surface Charge Considerations in the Pitting of Ion-Implanted Aluminum," *Journal of the Electrochemical Society*, vol. 135, no. 2, pp. 321–327, 1988.
[15] M. Vukasovich and J. Farr, "Molybdate in corrosion idnbition—a review," *Materials Performance*, vol. 25, pp. 9–18, 1986.
[16] M. Vukasovich and J. Farr, "Molybdate in corrosion inhibition—a review," *Polyhedron*, vol. 5, no. 1, pp. 551–559, 1986.
[17] K. Emregül and A. Aksüt, "The effect of sodium molybdate on the pitting corrosion of aluminum," *Corrosion science*, vol. 45, no. 11, pp. 2415–2433, 2003.
[18] E. Sherif and S.-M. Park, "Effects of 1, 5-naphthalenediol on aluminum corrosion as a corrosion inhibitor in 0.50 M NaCl," *Journal of The Electrochemical Society*, vol. 152, no. 6, pp. B205–B211, 2005.
[19] C. Monticelli, G. Brunoro, A. Frignani, and F. Zucchi, "Surface-active substances as inhibitors of localized corrosion of the aluminium alloy AA 6351," *Corrosion science*, vol. 32, no. 7, pp. 693–705, 1991.
[20] N. Ogurtsov, A. Pud, P. Kamarchik, and G. Shapoval, "Corrosion inhibition of aluminum alloy in chloride mediums by undoped and doped forms of polyaniline," *Synthetic metals*, vol. 143, no. 1, pp. 43–47, 2004.
[21] D. Zhu and W. van Ooij, "Corrosion protection of AA 2024-T3 by bis-[3-(triethoxysilyl) propyl] tetrasulfide in neutral sodium chloride solution. Part 1: corrosion of AA 2024-T3," *Corrosion Science*, vol. 45, no. 10, pp. 2163–2175, 2003.
[22] D. Zhu and W. van Ooij, "Corrosion protection of AA 2024-T3 by bis-[3-(triethoxysilyl) propyl] tetrasulfide in sodium chloride solution.: Part 2: mechanism for corrosion protection," *Corrosion Science*, vol. 45, no. 10, pp. 2177–2197, 2003.
[23] S. Doulami, K. Beligiannis, T. Dimogerontakis, V. Ninni, and I. Tsangaraki-Kaplanoglou, "The influence of some triphenylmethane compounds on the corrosion inhibition of aluminium," *Corrosion science*, vol. 46, no. 7, pp. 1765–1776, 2004.
[24] A. El-Etre, "Inhibition of acid corrosion of aluminum using vanillin," *Corrosion Science*, vol. 43, no. 6, pp. 1031–1039, 2001.
[25] H. Ashassi-Sorkhabi and S. Nabavi-Amri, "Corrosion inhibition of carbon steel in petroleum/water mixtures by N-containing compounds," *Acta Chimica Slovenica*, vol. 47, no. 4, pp. 507–518, 2000.
[26] E. Ebenso, U. Ekpe, B. Ita, O. Offiong, and U. Ibok, "Effect of molecular structure on the efficiency of amides and thiosemicarbazones used for corrosion inhibition of mild steel in hydrochloric acid," *Materials chemistry and physics*, vol. 60, no. 1, pp. 79–90, 1999.
[27] E. Ferreira, C. Giacomelli, F. Giacomelli, and A. Spinelli, "Evaluation of the inhibitor effect of L-ascorbic acid on the corrosion of mild steel," *Materials chemistry and physics*, vol. 83, no. 1, pp. 129–134, 2004.
[28] S. Saidman and J. Bessone, "Electrochemical preparation and characterisation of polypyrrole on aluminium in aqueous solution," *Journal of Electroanalytical Chemistry*, vol. 521, no. 1, pp. 87–94, 2002.
[29] O. Riggs, *Corrosion Inhibitors, Ed. Nathan, C.C*, ch. Theoretical Aspects of Corrosion Inhibitors and Inhibition, pp. 7–27. National Association of Corrosion Engineers—NACE, 1973.
[30] E. Banczek, S. Moraes, S. Assis, I. Costa, and A. Motheo, "Effect of surface treatments based on self-assembling molecules and cerium coatings on the AA3003 alloy corrosion resistance," *Materials and Corrosion*, vol. 64, pp. 199–206, 2011.

[31] N. Pirhady Tavandashti, S. Sanjabi, and T. Shahrabi, "Evolution of corrosion protection performance of hybrid silica based sol-gel nanocoatings by doping inorganic inhibitor," *Materials and Corrosion*, vol. 62, no. 5, pp. 411–415, 2011.
[32] A. Galio, S. Lamaka, M. Zheludkevich, L. Dick, I. Müller, and M. Ferreira, "Inhibitor-doped sol–gel coatings for corrosion protection of magnesium alloy AZ31," *Surface and Coatings Technology*, vol. 204, no. 9, pp. 1479–1486, 2010.
[33] D. Shchukin and H. Möhwald, "Smart nanocontainers as depot media for feedback active coatings," *Chemical Communications*, vol. 47, no. 31, pp. 8730–8739, 2011.
[34] M.-A. Chen, X.-B. Lu, Z.-H. Guo, and R. Huang, "Influence of hydrolysis time on the structure and corrosion protective performance of (3-mercaptopropyl) triethoxysilane film on copper," *Corrosion Science*, vol. 53, no. 9, pp. 2793–2802, 2011.
[35] A. Younis, *Protection of Aluminum Alloy (AA7075) from Corrosion by Sol-Gel Technique*. PhD thesis, Technische Universität Chemnitz, 2012.
[36] K. Ishizaki, S. Komarneni, and M. Nanko, *Porous materials: process technology and applications*. Kluwer Academic Publishers, 1998.
[37] F. Sittner and W. Ensinger, "Electrochemical investigation and characterization of thin-film porosity," *Thin solid films*, vol. 515, no. 11, pp. 4559–4564, 2007.
[38] G. Moretti and F. Guidi, "Tryptophan as copper corrosion inhibitor in 0.5 m aerated sulfuric acid," *Corrosion science*, vol. 44, no. 9, pp. 1995–2011, 2002.
[39] A. Younis, M. El-Sabbah, and R. Holze, "The effect of chloride concentration and pH on pitting corrosion of AA7075 aluminum alloy coated with phenyltrimethoxysilane," *Journal of Solid State Electrochemistry*, vol. 16, no. 3, pp. 1033–1040, 2012.
[40] M. Sheffer, A. Groysman, and D. Mandler, "Electrodeposition of sol–gel films on Al for corrosion protection," *Corrosion science*, vol. 45, no. 12, pp. 2893–2904, 2003.
[41] C. Higgins, D. Wencel, C. Burke, B. MacCraith, and C. McDonagh, "Novel hybrid optical sensor materials for in-breath O2 analysis," *Analyst*, vol. 133, no. 2, pp. 241–247, 2008.

Magnetic phenomena

Lecture Notes on Impedance Spectroscopy, Volume 4 – Kanoun (ed)
© 2014 Taylor & Francis Group, London, ISBN 978-1-138-00140-4

Precise eddy current measurements: Improving accuracy of determining of the electrical conductivity of metal plates

O. Märtens, R. Land & R. Gordon
Thomas Johann Seebeck Department of Electronics, Tallinn University of Technology, Estonia

M. Min
Thomas Johann Seebeck Department of Electronics, Tallinn University of Technology, Estonia
Competence Center ELIKO, Tallinn, Estonia

M. Rist
Competence Center ELIKO, Tallinn, Estonia

A. Pokatilov
AS Metrosert, Tallinn, Estonia
Department of Mechatronics, Tallinn University of Technology, Tallinn, Estonia

ABSTRACT: Eddy current (contactless) measurements, especially in the form of the impedance spectroscopy, could be effectively used to determine the electrical properties, including conductivity, of various materials, tissues and objects in many industrial, medical and other applications (e.g. coin validation). One important aspect is to have the calibration standards for the adjustment of eddy current instrumentation to guarantee repeatability and comparability of the measurements. In the paper, there is described an experimental study carried out to find the ways to improve and correct the theoretical eddy current models for the "single coil above the metal plate" setup, designed for measurements without using of the reference metal specimen with known electrical properties. The final results show, that the accuracy of measurement with a single-coil setup can be better than ±0.5% in the 60 to 500 kHz frequency range and ±3.0% at the frequencies up to 10 MHz, all valid for conductivity values from 2.5 to 25 MS/m.

Keywords: non-destructive testing, electrical conductivity, eddy current, calibration

1 INTRODUCTION

1.1 *About eddy current conductivity measurements*

Electrical resistivity (often expressed through the inverse value—conductivity) and its variations describe the properties of various metals and alloys [1], but also tissues and living bodies, characterized by the bio-impedance [2]. Electrical conductivity can be measured in the contact-less way by eddy current sensor(s), e.g. by using the measurement setup with a single measurement coil above the measured object, which can be a metal plate, for example, as described analytically [3] or numerically [4, 5]. Experiments show reasonable similarity between the real measurements and theoretical models.

Using of the vector (complex) measurement at multiple frequencies, which enables various penetration depths of electromagnetic fields into the material to be measured and can so give more information, allowing to determine beside the electrical conductivity also the liftoff value (distance between the measurement coil and the measured object). Also, the thickness of the material and even the thickness of various layers of the multi-layer structures can be determined. Liftoff calculation can be used to compensate the mechanical nonideality of the measurement setup [6, 7].

One important application field of eddy-current conductivity measurement is the recognition and validation of coins [8–10].

1.2 *Alternatives to single air-core coil measurement setup*

By idea, also two-coil alternatives to the "single coil" setup could be considered. In this case transmit and receive coils are on different sides of the metal plate [11] or the both coils are mounted "above the plate" [12].

Also, for metrological purposes, the NPL has proposed special configurations and shapes of both, the coil and reference metal specimen [13].

For precise absolute (not relative) measurements air-core coil solutions are used, while for the relative measurements, using of certain ferrite or other magnetic cores can be more efficient (SIGMASCOPE).

1.3 *About metrological aspects of measurements*

Accuracy of the eddy current conductivity measurements can be achieved by calibration of the instruments with known reference specimens (SIGMASCOPE). By opinion of the authors, obtaining or manufacturing of the metal pieces with precise and stable conductivity values in the wide frequency range (one-percent accuracy up to 500 kHz and few per cents within the 10 MHz range) is not an easy task, if possible at all.

More reasonable solution is to use analytical or numerical theoretical models [3–5, 11, 12] to obtain the conductivity value(s), from the measured complex impedance of the sensor-coil. Actually, the reverse models are needed here, as we can measure some kind of the transfer function of the electromagnetic sensors (coils), e.g. complex impedance at various frequencies.

Air-core solutions and models are more precise and stable than of the magnetic cores. The experimental work done so far, also shows that the single coil setup is not only simpler to model and calculate, but is giving also much better results in the mentioned frequency band. Still, our experiments show that for some specific and limited low frequency band (e.g. in the 10 kHz region) the "through-the-metal-plate" two-coil setup [11] can give better "absolute" measurement result.

Unfortunately, the solution, proposed by NPL [13] is working precisely in the very limited bandwidth up to 100 kHz.

One possible idea could be also the precise measurement of the direct current (DC) conductivity (or resistivity) of the metal plate with a 4-contact probe (precise instrumentation is available) and to use the found values as references for calibration of the eddy current instruments. Such comparison of the two methods has been carried out in [14], in which the eddy current method was tested up to 10 kHz frequency and obtained accuracy was characterized by about 3% error. Thus this method needs also further improvements to be used as measurement standards.

2 PROPOSED SOLUTION

2.1 *Starting point*

Single coil air-core measurement solutions have been investigated [14] (Märtens et al 2011) experimentally and compared with the theoretical models [3, 5] (Dodd and Deeds 1968, The odoulidis and Kotouzas 2000).

Also the need for having a reverse model offers the problem to be solved. We have to find the values of parameters of the measured material (object), while in the known works only the mathematical direct-models are given. Fortunately, this did not turn out to be a very complicated problem. The reverse-model has been implemented by using of the direct model in the calculation loop. We found out and used the gradients of changing of all the measurable and undefined parameters of the coil impedance and approached iteratively to the values to be defined. Special C/C++ software with direct and reverse models has been developed for PC based simulations. This software could be further embedded into micro-processor based devices. For experiments, the metal plates M1–M4 are used (80 × 80 × 3 mm size, see Fig. 1) with the following conductivities: M1–3.01 MS/m, M2–9.60 MS/m, M3–45.0 MS/m, M4–22.5 MS/m.

The PCB-based planar coil (a set of five coils is shown in Fig. 2) with N = 40 turns (diameter D = 12 mm) have been selected for further investigation. This type of coils has the most appropriate high accuracy frequency response up to 500 kHz and with a reasonable accuracy up to 10 MHz.

The experiments, together with applying the mentioned C/C++ reverse model to the measured complex impedance of the coil, give the significant decreasing of conductivity with increasing frequency (look the case of M2 in Fig. 3), being in the order of ten per cents in the frequency range 60–500 kHz and some tens of per cents for frequencies up to 10 MHz. As a result, it becomes evident that further corrections are obligatory to measure the conductivity more precisely.

2.2 *Further investigations done*

The mentioned single air-core coil measurement solution has been further investigated. Alternative copper-wire wounded coils with various numbers of turns have been tried. Also

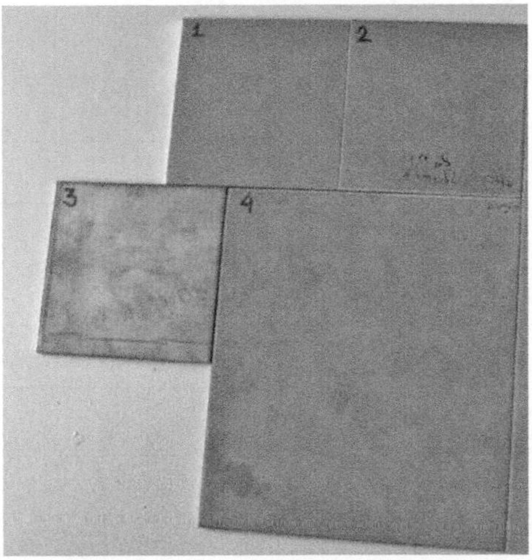
1
2
3
4

Figure 1. Used metal plates in experiments: M1–M4.

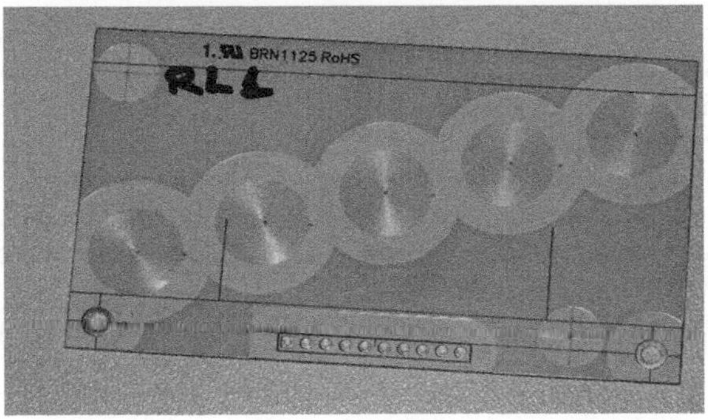
BRN1125 RoHS
RLL

Figure 2. A PCB-sensor with 5 coils, each with N = 40 turns and a diameter D = 12 mm.

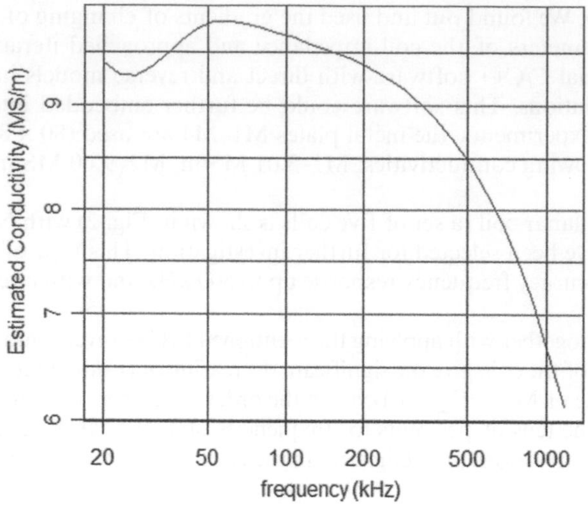
Estimated Conductivity (MS/m)
6
7
8
9
20
50
100
200
500
1000
frequency (kHz)

Figure 3. Frequency characteristics of the reversely estimated conductivity of M2.

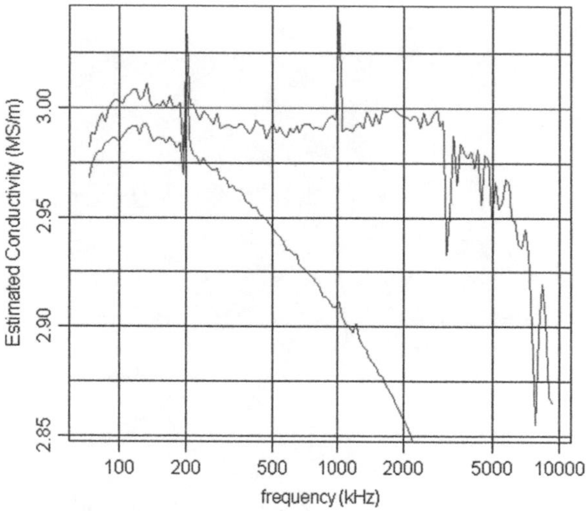
Estimated Conductivity (MS/m)
2.85
2.90
2.95
3.00
100
200
500
1000
2000
5000
10000
frequency (kHz)

Figure 4. Measured (lower curve) and the reverse-calculated (upper curve) conductivity of the plate M1.

two-coil setups with separate transmit and receive coils have been tried and compared with their theoretical models [11, 12] (Dodd and Deeds 1982, Burke and Ibrahim 2004). Anyway, the single-coil solution gave much better results in the frequency range under interest.

Alternatively, the experiments show that the "through the metal" [11] (Dodd and Deeds 1982) configuration can give even better absolute accuracy, but only at certain relatively low frequencies (e.g., at 10 kHz) and can be used, therefore, as a "reference" value for additional calibration and correction making for the single coil configuration. Nevertheless, this question needs further studies.

Various correction methods have been tried—primarily adding parasitic coil capacitance values to the model. An intuitive, but the most efficient correction was a simple approach, where the coil was first measured in the "air" (without any metal plate nearby). The difference of the impedance, measured at every single measurement frequency, from the ideal coil impedance $X_L = j\omega L$ with a zero value real part (ohmic resistance), was found and fixed

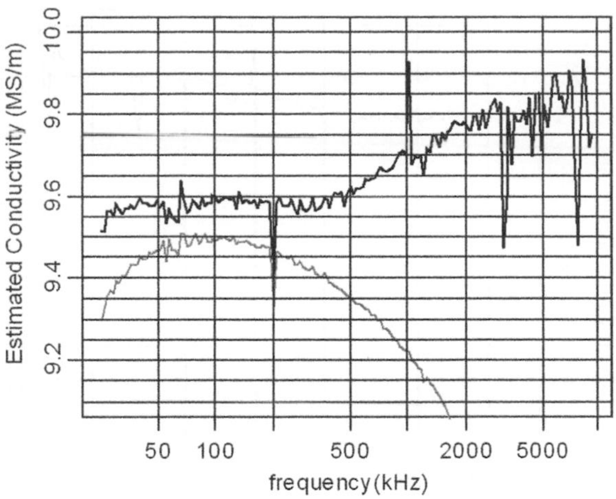
Estimated Conductivity (MS/m)
9.2
9.4
9.6
9.8
10.0
50
100
500
2000
5000
frequency(kHz)

Figure 5. Measured (lower curve) and reverse-calculated (upper curve) conductivity of the plate M2.

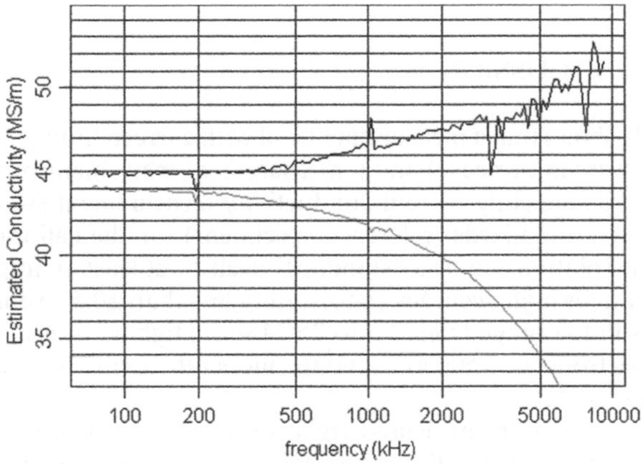
Estimated Conductivity (MS/m)
35
40
45
50
100
200
500
1000
2000
5000
10000
frequency (kHz)

Figure 6. Measured (lower curve) and reverse-calculated conductivity of the plate M3.

within a frequency range. Measurements showed that both—inductivity, and especially ohmic resistance of the coil impedance were slightly changing with the frequency. Our hypothesis is that the change of real part can be mainly explained by skin effect, where the effective cross-section area of wire is decreasing with the frequency rise. Changes in the inductive part can maybe explained by the influence of parasitic capacitances.

So, the correction of the complex impedance of the measurement coil has been found as a deviation from the impedance of an ideal one in the whole frequency range. The correction is simply an offset function.

Further, the correction of the complex impedance of the measurement coil has been found as a deviation from the impedance of an ideal one in the whole frequency range. This correction is a simple offset implementation via direct subtracting from the measured coil impedance with subsequent application of the reverse model (mentioned C/C++ software).

The results obtained this way show significant improvement in the measurement accuracy and give a very small frequency error in a wide range. The measured, compensated and "reverse calculated" results for the conductivity of metal plates from M1 to M4 are given in Figures 4 to 7 (upper curve for the corrected values and a lower line for raw non-corrected values).

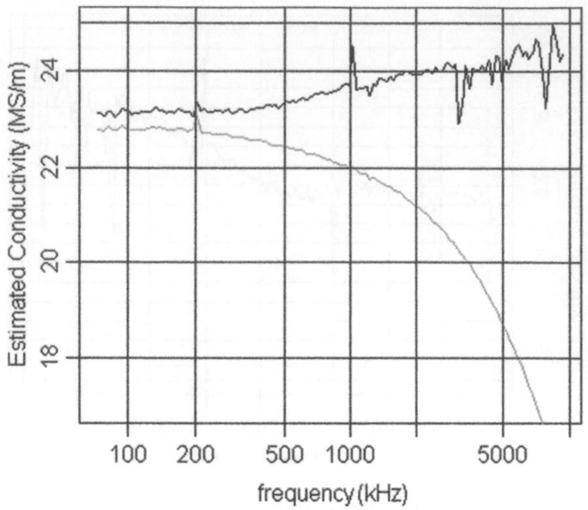
Estimated Conductivity (MS/m)
18
20
22
24
100
200
500
1000
5000
frequency (kHz)

Figure 7. Measured (lower curve) and reverse-calculated (upper curve) conductivity of the plate M4.

3 RESULTS AND DISCUSSION

The carried out study shows that a measurement coil with corresponding mathematical model (done as C/C++ software in our case) and a measurement instrument of the coil impedance can be used as an "absolute eddy current conductivity measurement system", not requiring any additional reference standards (reference specimens) for the calibration procedure. In other words, such the solution can be a "standard" itself or at least an important component in the metrology chain for high accuracy (uncertainty less than ±0.5%) eddy current conductivity measurements in the range from 2.5 to 25 MS/m. High accuracy measurement range covers the frequencies from 60 to 500 kHz and the moderate accuracy range (uncertainty less of ±3.0%) extends to 10 MHz.

Non-magnetic materials were measured (relative magnetic permeability $\mu = 1.0$) in the current research. In the future it will be reasonable to study also magnetic aspects of such measurements.

ACKNOWLEDGEMENT

Current work has been supported by the FP7-SME project "Safemetal", Enterprise Estonia (support to Competence Centre ELIKO), target financing project SF0140061s12 and the grant ETF8905 (Estonian Science Foundation). The research was supported also by European Union through the European Regional Development Fund. Special thanks to Dr. Theodoros P. Theodoulidis (University of Western Macedonia, Greece) for making available the effective eddy current models.

REFERENCES

[1] P. Rossiter, *The electrical resistivity of metals and alloys*. Cambridge University Press, 1991.
[2] O. Martinsen and S. Grimnes, *Bioimpedance and bioelectricity basics*. Academic press, 2011.
[3] C. Dodd and W. Deeds, "Analytical Solutions to Eddy-Current Probe-Coil Problems," *Journal of applied physics*, vol. 39, no. 6, pp. 2829–2838, 1968.
[4] C. Dodd and W. Deed, "Calculation of magnetic fields from time-varying currents in the presence of conductors," tech. rep., NASA STI/Recon, 1975.

[5] T.P. Theodoulidis and M. Kotouzas, "Eddy current testing simulation on a personal computer," in *Roma 2000 NDT World Conference*, 2000.
[6] P. Snyder, "Method and apparatus for reducing errors in eddy-current conductivity measurements due to lift-off by interpolating between a plurality of reference conductivity measurements," 1995.
[7] S. Linder, "Method and device for measuring the thickness and the electrical conductivity of electrically conducting sheets," 2010.
[8] A. Carlosena, A. Lopez-Martin, F. Arizti, A. Martínez-de Guerenu, J. Pina-Insausti, and J. García-Sayés, "Sensing in coin discriminators," in *Sensors Applications Symposium, 2007. SAS'07. IEEE*, pp. 1–6, IEEE, 2007.
[9] G. Howells, "Coin discriminators," 2009.
[10] J. Harris, J. Churchman, and D. Sharman, "Coin validation arrangement," 2007.
[11] C. Dodd and W. Deeds, "Absolute eddy-current measurement of electrical conductivity," tech. rep., Oak Ridge National Lab., TN (USA), 1981.
[12] S. Burke and M. Ibrahim, "Mutual impedance of air-cored coils above a conducting plate," *Journal of Physics D: Applied Physics*, vol. 37, no. 13, p. 1857, 2004.
[13] S. Harmon, M. Hall, L. Henderson, and P. Munday, "Calibration of commercial conductivity meters for measuring small items," *IEE Proceedings-Science, Measurement and Technology*, vol. 151, no. 5, pp. 376–380, 2004.
[14] N. Bowler and Y. Huang, "Electrical conductivity measurement of metal plates using broadband eddy-current and four-point methods," *Measurement Science and Technology*, vol. 16, no. 11, pp. 2193–2200, 2005.

Lecture Notes on Impedance Spectroscopy, Volume 4 – Kanoun (ed)

Wood fiber ferrite micro- and nano-composite materials for EMI-shielding

K. Dimitrov
University of Chemical Technology and Metallurgy Sofia, Bulgaria

T. Döhler, M. Herzog & S. Schrader
Technische Hochschule Wildau, Germany

S. Nenkova
University of Chemical Technology and Metallurgy Sofia, Bulgaria

ABSTRACT: A synthesis method for wood fiber ferrite micro- and nano-composites was developed and the properties of these materials were studied. The new method is based on a wood fiber modification by Fe^{+2}/Fe^{+3} aqueous solutions under formation of magnetite. The optimum synthesis conditions were defined and the modified wood fibers were used as filler in a polymer matrix to produce composite materials. Magnetite modified composites with special electromagnetic properties and microwave absorption ability were produced. Moreover, the raw material—wood is an economically feasible and sustainable source.

The new composite materials based on modified wood fibers and a polymer matrix (polystyrene-carbon black mixtures) were obtained via compression molding at 185°C. The composite materials are suitable for technical applications, e.g. in electromagnetic wave protection. The materials are characterized with respect to their morphological and electromagnetic properties. X-ray diffraction and SEM were used to investigate the wood fiber modifications. The dielectric permittivity, magnetic permeability and shielding efficient have been determined, in the frequency range from 1 MHz to 3 GHz.

Keywords: XRD, SEM, permittivity, permeability, shielding effect

1 INTRODUCTION

Large amounts of magnetic particles have been produced for magnetic recording media in the last fifty years, while lately, magnetic particles have also been extensively used in a variety of applications such as ferrofluids, magnetic inks, for means of extraction and purification of biological materials, and clinical diagnostics [1]. Magnetite (Fe_3O_4) particles represents a very interesting type of magnetic materials, being used among others in information storage, magnetic fluids and in medical applications (i.e. drug carriers, imaging agents), as well as in materials for absorption of electromagnetic radiation. Such a wide variety of possible uses and application have warranted much attention and results in a large body of research work concerning the synthesis of magnetite.

Composite materials are of interest due to the potential synergistic properties that may arise from the combination of two or more precursors. Two such precursors are wood or cellulose fibers and magnetic nanoparticles. These hybrid materials exhibit the inherent properties of the fiber substrate, in particular flexibility and strength, and also the magnetic properties of the surface bonded nanoparticles. Superparamagnetic Iron Oxide Nanoparticles (SPION) covered with a polymer have been used in medical research such as devices for cell isolation, immobilization of enzymes, controlled releasesystems and separation of biological materials [2]. Cellulose beads covering inorganic particles, a new kind of composite materials, can be used as organic support. Guo et al. reported the study of inorganic adsorbents inside cellulose beads in the elimination of arsenic from aqueous solutions [3]. Thus, cellulose beads could act as a column bed, suitable for liquid chromatography. Another way of using cellulose beads is to place a magnetic particle inside the structure and thus the new material could react as a magnetic actuator. Moreover, the synthesis of the magnetic cellulose [4] or polysaccharide like carboxymethyl-cellulose [5] and carrageenan [6] has generated a great attention in recent years due to their specific properties: non-toxic nature, biocompatibility, variable ionic permeability, combined with a high hydrophilic effect and considerable mechanical strength. The immobilization of α-amylase onto cellulose-coated magnetite nanoparticles has been reported in the starch degradation study [7]. Previous work describes the synthesis of cellulose beads bearing micrometric magnetite particles and papain immobilization study [8]. In this initial report, the covered micrometric magnetite in a cellulosic matrix results in a net reducing effect of magnetic coercive field with consequential formation of a superparamagnetic cellulose composite.

Magnetically responsive cellulose fibers will allow the investigation of new concepts in papermaking and packaging, security paper, and information storage. Potential applications are in electromagnetic shielding, magnetographic printing and magnetic filtering [4, 9, 10].

Throughout the literature, there have been reports about superparamagnetic papers obtained through "in situ" synthesis of ferrites [9–11] and others concerning ferromagnetic paper obtained by the "lumen-loading" approach, whereby the lumen of the cellulose fiber is loaded with commercial pigments such as magnetite and maghemite.

In this study, wood fibers modified with nano-sized magnetite have been developed. They show improved electrical conductivity and microwave adsorption ability. Based on these fibers new polymer composites, with special shielding properties against electromagnetic wave,have been obtained.

For this purpose wood fibers have been modified by, Fe^{+2} and Fe^{+3}, in order to obtain composite materials with the desired properties.

2 SYNTHESIS

For synthesis of the composite materials, following reagents were used: bonding agent polystyrene with 5% carbon black. Carbon black was received from TIMCAL Graphite & Carbon Belgium. This product was used to produce material with enhanced electrical conductivity. As second filler wood fibers from hard wood with a length between 2 mm and 20 mm, were used. Their diameters were below 500 μm. The modifying agents were $FeCl_2 \cdot 4H_2O$, $FeCl_3 \cdot 6H_2O$ and 30 wt.% NH_3 solution.

3 WOOD FIBERS MODIFICATION

Wood fibers (500 g) were treated with a supension of magnetite. To produce magnetite, we used solutions of $FeCl_2 \cdot 4H_2O$ (59.64 g in 300 ml distilled water) and $FeCl_3 \cdot 6H_2O$ (162.46 g in 600 ml distilled water). This solution was treated with 30 wt.% NH_3 solution and the obtained black suspension is vigorously stirred for 20 min. Fe_3O_4 precipitate was decanted. Wood fibers were mixed with Fe_3O_4 precipitate to full absorption of magnetite particles. Modified fibers were dried for 48 hours in air at 37°C. Magnetite was produced following the reaction: $FeCl_2 + 2FeCl_3 + 8NH_4^+ + 8OH^- \rightarrow Fe_3O_4 \downarrow + 8Cl^- + 4H_2O + 8NH_4^+$ [12].

4 PREPARATION OF BONDING AGENT

Polystyrene and Carbon black were used for preparation of the bonding agent. Polystyrene and Carbon black were mixed in a BarabenderPlasti-Corder® Lab Station, in the attachment Mixer 350 E at 190°C. The rotation speeds of the blades were 20 rpm.

Table 1. Formulation of composite materials.

Samples	Amount PS + 5% CB (%)	Amount mod. WF (%)	Weight of samples (g)
PSCB-WF(1)	70	30	200
PSCB-WF(2)	60	40	200
PSCB-WF(3)	50	50	200

5 PREPARATION OF COMPOSITE MATERIALS

The composite synthesis follows a simple physical mixing of wood fibers and bonding agent (polystyrene + carbon black). Mixing is carried out in a Barabender Plasti-Corder® Lab Station, in the attachment Mixer 350 E. A defined amount of the components was placed in the mixer and treated by following conditions: the temperature was 190°C, the rotation speed of the blades was 20 rpm. Then the material is pressed (10 MPa) in a laboratory press Servitec/Polystat 300 S. Table 1 shows the formulation for different cases with the following abbriviations: PS—polystyrene, CB—carbon black, WF—wood fiber.

6 ANALYTICAL METHODS

6.1 *X-ray Diffraction (XRD)*

XRD investigations of the composites were performed using a D8-Advance from Bruker AXS, equipped with a Ge (111)—Johansson monochromator. The data were collected in the 2θ range from 10° to 80°, with a step width of 0.01°.

6.2 *Scanning Electron Microscopy (SEM)*

For the SEM investigation, specimens with surface areas of about 5 × 5 mm were taken from the modified wood fibers. The samples were coated with gold by argon sputtering. Subsequently surface morphology of the materials were investigated by SEM—JSM 6400.

6.3 *Electromagnetic properties*

The electrical impedance, conductivity, real and imaginary part of dielectric permittivity and magnetic permeability have been determined using an Agilent E4991A RF Spectrum Analyzer. The used frequency was in the range from 1 MHz to 3 GHz. The sample thickness varied from 0.5 to 1 mm. The frequency step was 20 Hz. The tests were performed at room temperature (≈25°C)

7 RESULTS AND DISCUSSIONS

7.1 *Characterization of chemical modified wood fibers*

Figure 1 presents results of a X-ray spectrum from the material powder. The reflections of 14.8°, 16.6°, 22.4° correspond to the crystal line part of cellulose in wood fibers. The broader reflection at 30°, 35.8°, 57.3°, 62.8° are typical for the magnetite phase and also the reflection is caused by the spinell structure of Fe_3O_4.

Figure 2, ((a), (b), (c)) presents SEM micrographs of chemical modified wood fibers with different resolution. These figures show successful modification of wood fibers. There are magnetite particles with different size. This is because, part of magnetite particles agglomerate during the modificaion process.

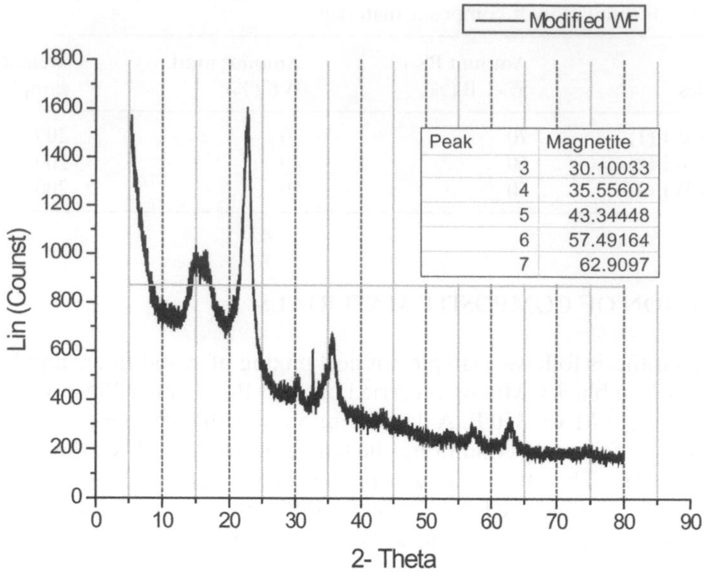
Modified WF
Peak | Magnetite
3 | 30.10033
4 | 35.55602
5 | 43.34448
6 | 57.49164
7 | 62.9097
Lin (Counst)
2- Theta

Figure 1. X-ray diffraction of modified wood fibers.

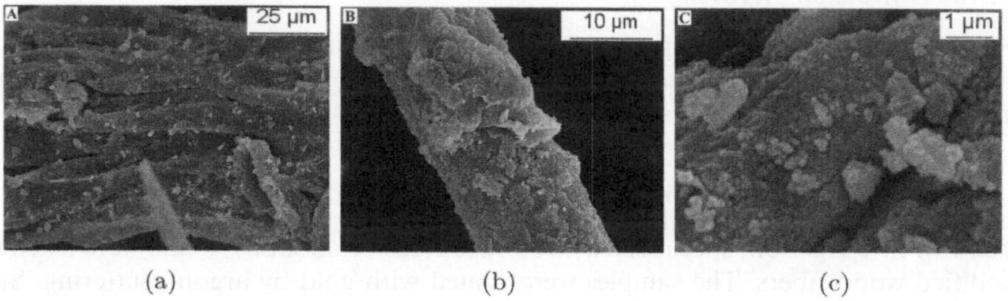
A
25 μm
B
10 μm
C
1 μm
(a)
(b)
(c)

Figure 2. SEM of modified wood fibers.

7.2 *Electrical characterization*

Conducting polymer composites should have a high dielectric constant and a high dissipation factor in the relevant frequency regions if they are used in charge storing devices, decoupling capacitors and electromagnetic interference shielding applications. For studying the dielectric properties, the capacitance and dissipation factor of the samples were measured in the frequency range 1 MHz–3 GHz, using an Agilent E4991A RF Spectrum Analyzer. Samples have been prepared from plane-parallel composite plates. Their thicknesses were ranging from 1 mm to 1.5 mm. Metal electrodes have been applied to both sides of the plates in guard-ring configuration. The electrode diameter was 10 mm.

Figure 3 shows the real part ε' and the imaginary part ε'' of the complex dielectric constant $\varepsilon = \varepsilon' - j\varepsilon''$ of conducting polymer composites in dependence on frequency. The permittivity measurements show resonance polarization of the composite materials (Fig. 3). The resonance frequency is around 1.5 GHz. This could be explained with polarization and conductivity losses of carbon black particles in the polymer matrix. The measurements show low relative permittivity almost in the whole range of frequencies. The composites possess high values of ε', ε'' and high $\tan\delta$ in the radio frequency range, at 1.5 GHz indicating that the composites could be utilized as EMI shielding material at this frequency range. Figure 4 shows the loss factor $\tan\delta = \varepsilon''/\varepsilon'$ versus frequency for selected composites.

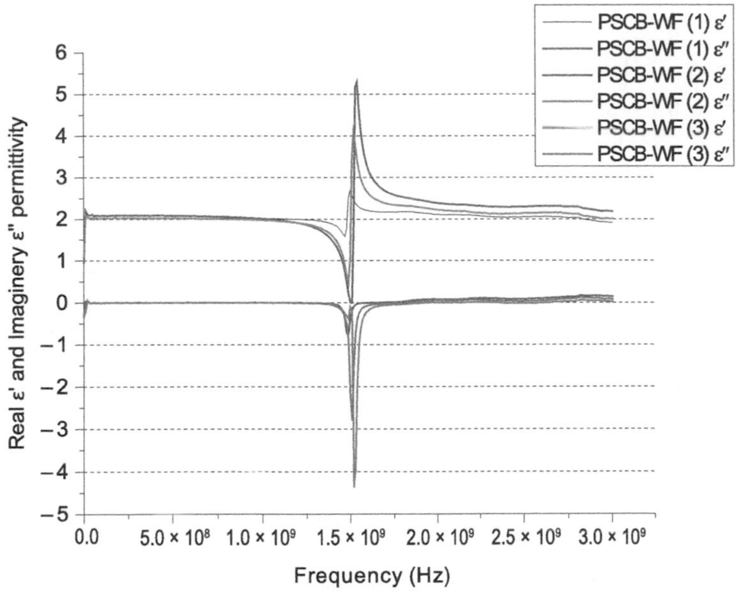
PSCB-WF (1) ε'
PSCB-WF (1) ε"
PSCB-WF (2) ε'
PSCB-WF (2) ε"
PSCB-WF (3) ε'
PSCB-WF (3) ε"
Real ε' and Imaginery ε" permittivity
Frequency (Hz)
6
5
4
3
2
1
0
−1
−2
−3
−4
−5
0.0
5.0 × 10⁸
1.0 × 10⁹
1.5 × 10⁹
2.0 × 10⁹
2.5 × 10⁹
3.0 × 10⁹

Figure 3. Variation of (ε') and (ε'') of the composites as a function of frequency.

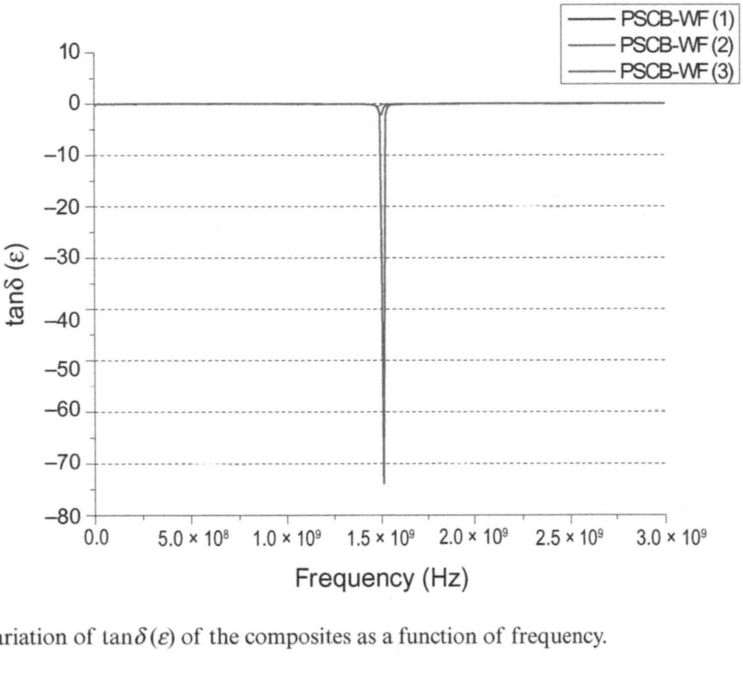
PSCB-WF (1)
PSCB-WF (2)
PSCB-WF (3)
tanδ (ε)
Frequency (Hz)
10
0
−10
−20
−30
−40
−50
−60
−70
−80
0.0
5.0 × 10⁸
1.0 × 10⁹
1.5 × 10⁹
2.0 × 10⁹
2.5 × 10⁹
3.0 × 10⁹

Figure 4. Variation of $\tan\delta(\varepsilon)$ of the composites as a function of frequency.

The permeability measurement (Fig. 5) show magnetic properties of the composites as a function of frequency. Materials used for EMI shielding should have a high loss number (μ''). The measurement shows a high imaginary permeability until 100 MHz. This is typical for ferromagnetic spinel. Near 2 GHz, a weak resonance can be observed. The strength of this process increases with the addition of nano-particles, but the resonance frequency seems not to depend on the particle concentration. That is what also theoretical models predict for single-domain particles with an isotropic distribution of magnetic orientations.

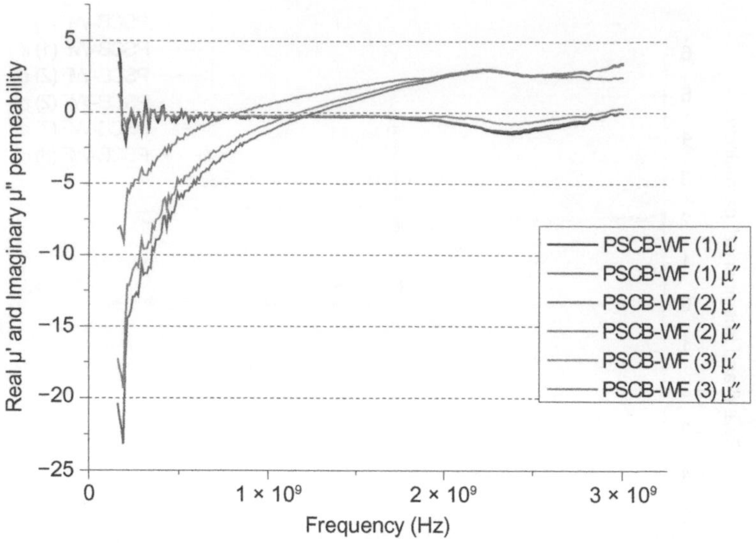

5
0
−5
−10
−15
−20
−25
Real μ' and Imaginary μ" permeability
0
1 × 10⁹
2 × 10⁹
3 × 10⁹
Frequency (Hz)
PSCB-WF (1) μ′
PSCB-WF (1) μ″
PSCB-WF (2) μ′
PSCB-WF (2) μ″
PSCB-WF (3) μ′
PSCB-WF (3) μ″

Figure 5. Variation of permeability (real μ′and imaginary μ″ part) as a function of frequency.

7.3 *EMI shielding*

For a transverse electromagnetic wave propagating in a sample with low magnetic interaction, the total Shielding Efficiency (SE_T) of the sample is expressed as Eq. (1) [13–15]:

$$SE_T = \log 10\left(\frac{P_{in}}{P_{out}}\right) = SE_A + SE_R \tag{1}$$

where P_{in} and P_{out} are the power incident on and transmitted through a shielding material, respectively. The SE_T is expressed in decibel (dB). SE_A is the absorption and SE_R the reflection from both sides of the material neglecting multiple reflections inside the shielding materials. The terms in Eq. (1) can be described as:

$$SE_A = 8.86\alpha l \tag{2}$$

$$SE_R = 20\log\left(\frac{|1+n|^2}{4|n|}\right) \tag{3}$$

where the parameters α and n are defined by the following equations, l is the thickness of the shielding barrier.

$$\alpha = \frac{2\pi}{\lambda}\sqrt{\frac{\varepsilon'\sqrt{1+\tan\delta^2}}{2}} \tag{4}$$

$$n = \sqrt{\frac{\varepsilon'(\sqrt{1+\tan\delta^2}+1)}{2}} - i\sqrt{\frac{\varepsilon'(\sqrt{1+\tan\delta^2}-1)}{2}} \tag{5}$$

where λ is the wave length, ε' the real part of the complex relative permittivity.

Using equations 4 and 5 values of SE_R and SE_A for the PSCB-WF composites were calculated. The SE_R and SE_A of the composites as a function of frequency (1 MH–3 GHz) are shown in Figs. 6 and 7, respectively. A high value (around 40 dB) of SE_R is obtained at 1.5 GHz for the composite PSCB-WF(2) because the composite has high values of ε' and $\tan\delta$. The SE_A of composites increases with frequency because SE_A is directly proportional to the frequency via eqns. 2 and 4. The maximum value (4 dB) of SE_A was obtained at 3 GHz for all composites.

The total SE_T of the composites was calculated by adding SE_R and SE_A. The SE_T of the composites as a function of frequency is shows in Fig. 8.

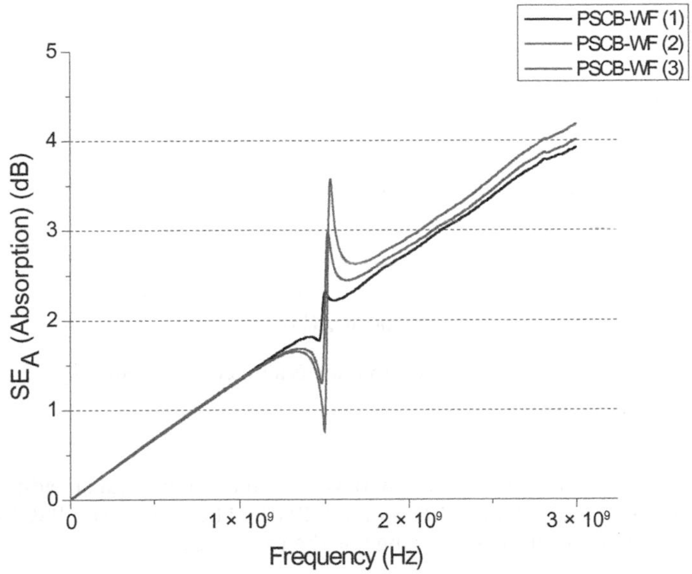
PSCB-WF (1)
PSCB-WF (2)
PSCB-WF (3)
5
4
3
2
1
0
SE_A (Absorption) (dB)
0
1 × 10⁹
2 × 10⁹
3 × 10⁹
Frequency (Hz)

Figure 6. Variation of SE_A for PSCB-WF composites as a function of frequency.

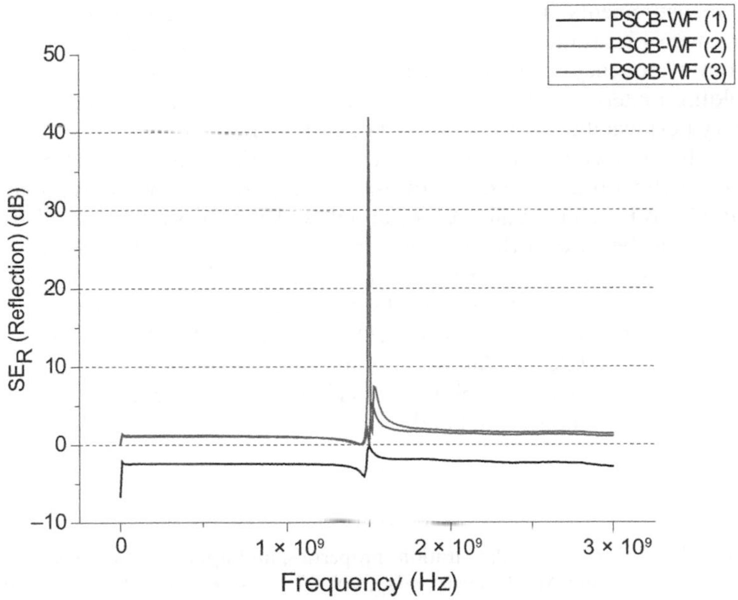
PSCB-WF (1)
PSCB-WF (2)
PSCB-WF (3)
50
40
30
20
10
0
−10
SE_R (Reflection) (dB)
0
1 × 10⁹
2 × 10⁹
3 × 10⁹
Frequency (Hz)

Figure 7. Variation of SE_R for PSCB-WF composites as a function of frequency.

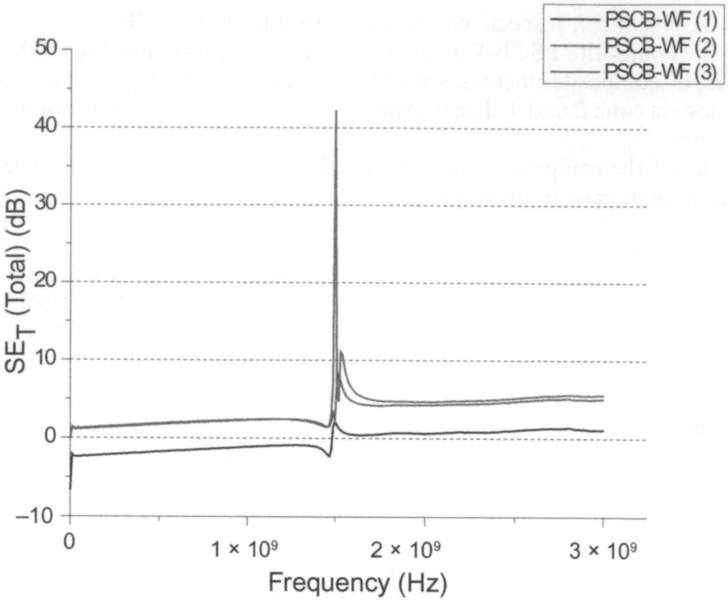
PSCB-WF (1)
PSCB-WF (2)
PSCB-WF (3)
50
40
30
20
10
0
−10
SE_T (Total) (dB)
0
1 × 10⁹
2 × 10⁹
3 × 10⁹
Frequency (Hz)

Figure 8. Variation of SE_T for PSCB-WF composites as a function of frequency.

The result shows that, the composite PSCB-WF (2) has the highest shielding effect round 1.5 GHz. In this range of frequency, *SE* is about 40 dB. The Composite PSCB-WF(3) has a smaller *SE* of about 10 dB in the same range of the frequency.

8 CONCLUSION

XRD and SEM investigation show successful magnetite modification of wood fibers. The permittivity measurements show resonance polarization of the composite materials. The resonance frequency is around 1.5 GHz. The composites possess high values of ε' and high $\tan\delta$ in the radiofrequency range, at 1.5 GHz indicating that the composites could be utilized as EMI shielding material at this frequency range. The permeability measurement shows a high imaginary permeability until 100 MHz The composites could be utilized as EMI shielding material at this frequency range. The result show that the composite PSCB-WF(2) has the highest shielding effect round 1.5 GHz. In this range of frequency, *SE* is about 40 dB. The Composite PSCB-WF(3) has smaller *SE* of about 10 dB in the same range of frequency.

The Material can be used in the frequency range from 1427 to 1535 MHz, the 1.5 GHz Band. Currently it is used in a number of radio communication services, including: point to point fixed services, point to multipoint fixed services in rural and remote areas for the operation of Digital Radio Concentrator Systems. (DRCS) used for the delivery of public telecommunications services, mobile-satellite services and aeronautical mobile telemetry services. The investigated materials can be exploited for effective EMI-Shielding against radiation of this extensively used communication band.

REFERENCES

[1] M. Ozaki, "Magnetic particles: Preparation, properties, and applications," in *Surface and Colloid Science* (E. Matijević and M. Borkovec, eds.), vol. 17 of *Surface and Colloid Science*, pp. 1–26, Springer US, 2004.

[2] J. Ramírez-Vick, A.A. García, and J. Lee, "Immobilization of silver ions onto paramagnetic particles for binding and release of a biotin-labeled oligonucleotide," *Reactive and Functional Polymers*, vol. 43, no. 1, pp. 53–62, 2000.

[3] X. Guo, Y. Du, F. Chen, H.-S. Park, and Y. Xie, "Mechanism of removal of arsenic by bead cellulose loaded with iron oxyhydroxide (<i>β</i>-feooh): Exafs study," *Journal of colloid and interface science*, vol. 314, no. 2, pp. 427–433, 2007.

[4] A.C. Small and J.H. Johnston, "Novel hybrid materials of magnetic nanoparticles and cellulose fibers," *Journal of colloid and interface science*, vol. 331, no. 1, pp. 122–126, 2009.

[5] P. Sipos, "Manufacturing of size controlled magnetite nanoparticles potentially suitable for the preparation of aqueous magnetic fluids," *Romanian reports in physics*, vol. 58, no. 3, p. 269, 2006.

[6] A.L. Daniel-da Silva, T. Trindade, B.J. Goodfellow, B.F. Costa, R.N. Correia, and A.M. Gil, "In situ synthesis of magnetite nanoparticles in carrageenan gels," *Biomacromolecules*, vol. 8, no. 8, pp. 2350–2357, 2007.

[7] M. Namdeo and S. Bajpai, "Immobilization of *α*-amylase onto cellulose-coated magnetite (ccm) nanoparticles and preliminary starch degradation study," *Journal of Molecular Catalysis B: Enzymatic*, vol. 59, no. 1, pp. 134–139, 2009.

[8] J. Correa, D. Canetti, E. Bordallo, J. Rieumont, and J. Dufour, "Application of cubic magnetite to the synthesis of super paramagnetic cellulose beads for enzyme immobilization," in *Proceedings of the 8th Inter American Congress of Electron Microscopy, Inter American Committee of Societies of Electron Microscopy, Rev. Biotecnol. Aplic.*, 2005.

[9] R. Marchessault, P. Rioux, and L. Raymond, "Magnetic cellulose fibres and paper: Preparation, processing and properties," *Polymer*, vol. 33, no. 19, pp. 4024–4028, 1992.

[10] L. Raymond, J.-F. Revol, D. Ryan, and R. Marchessault, "In situ synthesis of ferrites in cellulosics," *Chemistry of Materials*, vol. 6, no. 2, pp. 249–255, 1994.

[11] J.A. Carrazana-García, M. Lopez-Quintela, and J. Rivas-Rey, "Characterization of ferrite particles synthesized in presence of cellulose fibers," *Colloids and Surfaces A: Physicochemical and Engineering Aspects*, vol. 121, no. 1, pp. 61–66, 1997.

[12] H.-D. Hunger and K. Vaskova, "Präparation und Charakterisierung von biologisch aktiven Magnetit-Protein-Nanopartikeln," tech. rep., TH Wildau, 2006.

[13] Z. Liu, G. Bai, Y. Huang, Y. Ma, F. Du, F. Li, T. Guo, and Y. Chen, "Reflection and absorption contributions to the electromagnetic interference shielding of single-walled carbon nanotube/polyurethane composites," *Carbon*, vol. 45, no. 4, pp. 821–827, 2007.

[14] J. Joo and A. Epstein, "Electromagnetic radiation shielding by intrinsically conducting polymers," *Applied physics letters*, vol. 65, no. 18, pp. 2278–2280, 1994.

[15] R.B. Schulz, V. Plantz, and D. Brush, "Shielding theory and practice," *Electromagnetic Compatibility, IEEE Transactions on*, vol. 30, no. 3, pp. 187–201, 1988.

Lecture Notes on Impedance Spectroscopy, Volume 4 – Kanoun (ed)
© 2014 Taylor & Francis Group, London, ISBN 978-1-138-00140-4

Author index

Printed and bound by CPI Group (UK) Ltd, Croydon, CR0 4YY

Printed and bound by CPI Group (UK) Ltd, Croydon, CR0 4YY
18/10/2024
01776251-0006